Continua, Decompositions, Manifolds

Continua
Decompositions
Manifolds

Proceedings of Texas Topology Symposium 1980

Edited by R. H. Bing, William T. Eaton, and Michael P. Starbird

 UNIVERSITY OF TEXAS PRESS, AUSTIN

International Standard Book Number 0-292-78061-3
Library of Congress Catalog Card Number 83-80310
Copyright © 1983 by the University of Texas Press
All rights reserved
Printed in the United States of America

First Edition, 1983

Requests for permission to reproduce material from this
work should be sent to Permissions, University of Texas
Press, Box 7819, Austin, Texas 78712.

For reasons of economy and speed, this volume has been
printed from camera-ready copy furnished by the editors,
who assume full responsibility for its contents.

Contents

Preface   vii

Contributors   ix

Continua Theory

Some Topics in Modern Continua Theory
   David P. Bellamy   1

Stable Homeomorphisms, Galois Spaces, and Related Properties
in Homogeneous Continua
   Wayne Lewis and Judy Kennedy Phelps   27

Principal Embeddings of Atriodic Plane Continua
   John C. Mayer   34

When Homogeneous Continua Are Hausdorff Circles
(or Yes, We Hausdorff Bananas)
   Forest W. Simmons   62

A Classification of Certain Inverse Limit Spaces
   Will Watkins   74

A Note Concerning a Continuum of J. C. Mayer
   Sam W. Young   83

On the Relative Complexity of a Tree-like Continuum
and Its Proper Subcontinua
   Sam W. Young   85

Decomposition Spaces

Saturated 2-Sphere Boundaries in Bing's Straight-Line
Segment Example
   Steve Armentrout   96

Decompositions of S^3 into Circles
   D. S. Coram   111

A Mismatch Property in Spherical Decomposition Spaces
   Robert J. Daverman   119

Countable Starlike Decompositions of S^3
   Richard Denman   128

General Position Properties Related to the Disjoint Discs Property
   Dennis J. Garity   132

A Discussion of Results and Problems Related to Cellularity in Polyhedra
James P. Henderson   141

Raising the Dimension of 0-Dimensional Decompositions of E^3
Louis F. McAuley and Edythe P. Woodruff   152

Manifolds

Curves Isotopic to Tame Curves
M. Brin   163

Symplectic Maps on Open Cells: A Fixed Point Theorem
M. R. Colvin and K. Morrison   167

A Structure Set Analogue of Chapman-Ferry-Quinn Theory
F. T. Farrell and W. C. Hsiang   172

Heegaard Splittings of Homology Three-Spheres and Homotopy Three-Spheres
D. R. McMillan, Jr.   191

Linear Isotopies and Miscellaneous Topics

Shape Properties of Compacta in Generalized n-Manifolds
S. Armentrout and S. Singh   202

On Almost Continuous Functions
Marwan M. Awartani and Samir A. Khabbaz   221

On the Problems Related to Linear Homeomorphisms, Embeddings, and Isotopies
Robert Connelly, David W. Henderson, Chung-wu Ho, and Michael Starbird   229

Simplicial Complexes Homeomorphic to Proper Self-Subsets Have Free Faces
David W. Henderson   240

The Alexander Linear Isotopy Theorem in E^3
Michael Starbird   243

Preface

During the summer of 1980, The University of Texas at Austin hosted a six-week conference on topology supported by the National Science Foundation, The University of Texas, and individual participants. During the first two weeks, Professor David Bellamy gave a series of lectures on continua theory. Professor W. C. Hsiang lectured during the second two-week period on approximating homotopy equivalences by homeomorphisms, and Professor Robert Daverman lectured on decomposition spaces during the last two weeks of the conference.

Afternoons were devoted to contributed talks by other conference participants and water-skiing seminars at the local lakes. The conference was stimulating mathematically and enjoyed by all. The University of Texas at Austin intends to host similar summer seminars in the future.

We would like to thank the National Science Foundation and The University of Texas for their generous support. We would especially like to thank all the participants whose activity and enthusiasm made the conference such a success. We also thank Colleen Kieke for typing these proceedings, and Richard Denman and Richard Skora for help in proofreading.

R. H. BING
WILLIAM T. EATON
MICHAEL P. STARBIRD

Contributors

Steve Armentrout
Pennsylvania State University
University Park, Pennsylvania
16802

Marwan M. Awartani
Birzeit University
Birzeit, West Bank
Israel

David P. Bellamy
University of Delaware
Newark, Delaware 19711

Matthew Brin
SUNY Binghamton
Binghamton, New York 13901

M. R. Colvin
University of Alabama
University, Alabama 35486

Robert Connelly
Cornell University
Ithaca, New York 14853

D. S. Coram
Oklahoma State University
Stillwater, Oklahoma 74074

Robert J. Daverman
University of Tennessee
Knoxville, Tennessee 37916

Richard Denman
Southwestern University
Georgetown, Texas 78626

F. T. Farrell
University of Michigan
Ann Arbor, Michigan 48109

Dennis J. Garity
Oregon State University
Corvallis, Oregon 97331

David W. Henderson
Cornell University
Ithaca, New York 14853

James P. Henderson
Texas A & M University
College Station, Texas 77843

W. C. Hsiang
Princeton University
Princeton, New Jersey 08544

Chung-wu Ho
Southern Illinois University
Edwardsville, Illinois 62026

Samir A. Khabbaz
Lehigh University
Bethlehem, Pennsylvania 18015

Wayne Lewis
Texas Tech University
Lubbock, Texas 79409

John C. Mayer
University of Saskatchewan
Saskatoon, Saskatchewan S7N 0W0
Canada

Louis F. McAuley
SUNY Binghamton
Binghamton, New York 13901

D. R. McMillan, Jr.
University of Wisconsin
Madison, Wisconsin 53706

K. Morrison
California Polytechnic
   State University
San Luis Obispo, California 93407

Judy Kennedy Phelps
Auburn University
Auburn, Alabama 36830

Forest W. Simmons
North Texas State University
Denton, Texas 76203

S. Singh
Southwest Texas State University
San Marcos, Texas 78666

Michael P. Starbird
University of Texas at Austin
Austin, Texas 78712

Will Watkins
Pan American University
Edinburg, Texas 78539

Edythe P. Woodruff
11 Fairview Ave.
East Brunswick, New Jersey 08816

Sam W. Young
Auburn University
Auburn, Alabama 36830

Continua, Decompositions, Manifolds

Some Topics in Modern Continua Theory

David P. Bellamy

Continua theory is easily defined as the study of compact, connected, Hausdorff spaces, with attention usually being restricted to metric spaces. It has traditionally come to mean the study of such spaces with "bad" properties, either local or global, such as non-local connectedness or indecomposability. The purpose of this article is to give a brief overview of recent work in continua theory, hopefully useful to the specialist and informative to other mathematicians. There are many interesting unsolved problems relating to the topics considered. These are deferred to the problem section and are not discussed at length in the text. None of the results contained herein are mine except those referenced as such or claimed as new or original, though I am uncertain about the originality of Theorem 2 of Section IV.

The word continuum is used here to mean a compact connected Hausdorff space; if metrizability is assumed, as will frequently be the case, it will be explicitly stated. F. B. Jones has often described the class of continua as a spectrum, ranging from the hereditarily indecomposable and indecomposable continua on the left, through a vast and poorly undertstood intermediate area of progressively less pathological continua, to Peano continua, finite polyhedra, and finally compact manifolds on the right.

The intermediate area is difficult to study systematically because of the extreme diversity of the known examples. A few ideas, such as homogeneity, run across the whole spectrum like threads. However, for studying the intermediate area *per se*, one of the most useful concepts ever developed is Jones' notion

of aposyndesis. This is the point of departure for the first
section of this article.

I. SET FUNCTIONS AND APOSYNDESIS

The concept of aposyndesis has led to the development and
study of several set-valued set functions. I shall discuss
one of these at some length and mention others.

Let S be a continuum and let $A \subseteq S$. $T(A)$ is the set of
points having no closed, connected neighborhood which misses
A. Such neighborhoods are hereinafter called *continuum
neighborhoods*. $T(p)$ is written for $T(\{p\})$. The set function T has properties somewhat akin to those of the closure
operator, as might be expected. The principal differences,
formally, are that it may happen that $T(A) \neq T(T(A))$ or
$T(A \cup B) \neq T(A) \cup T(B)$ for $A, B \subseteq S$.

Some examples of continua may be helpful in visualizing
the set function T geometrically.

Example la: Let X denote the cone over the Cantor set, with
vertex v, base B, and r some point not in $B \cup \{v\}$.

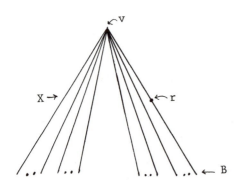

Fig. 1.

Here, $T(v) = X$; $T(r)$ is the segment down from   r   to   B,
while $T(B) = B$. For this example, $T(A) = \cup\{T(p) \mid p \in A\}$ for
any closed $A \subseteq X$.

Example 1b: If the vertex, v_2,   of another copy, X_2 of   X
is attached to some point of the base of   X   to form a new
continuum   Y,   then $T_Y^2(v) = T_Y(T_Y(v)) = X \cup X_2 = Y$.

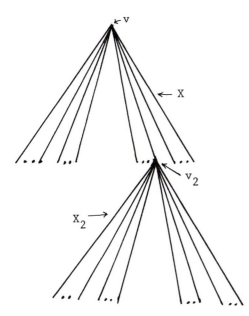

Fig. 2.

(T(A)   is, of course, computed with respect to the topology
of whatever continuum   A   is being considered as a subset of.
If confusion is possible, the name of the continuum will be
used as a subscript on   T.)

<u>Example 2</u>: Let Z denote the closure in the plane of the
graph $y = \sin \dfrac{1}{x}$ for $0 < x \le \dfrac{1}{\pi}$.

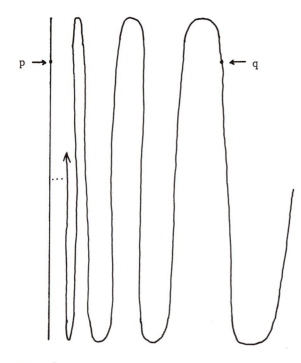

$p \rightarrow$ $\leftarrow q$

Fig. 3.

Here, $T(p)$ is the entire vertical limit arc while
$T(q) = \{q\}$. For this example $T^2 = T$.

<u>Example 3</u>: Let M denote the suspension of the Cantor set.
Here $T(p) = p$ for every $p \in M$, but if u and v are the
vertices, $T(\{u,v\}) = M$.

T   can be used to unify the study of several properties of continua. For instance, we have the following observations, mostly due to F. B. Jones or H. S. Davis. They can be taken as the definitions of the properties considered.

   a) S is *connected im Kleinen* at $p \in S$ if and only if for each closed $A \subseteq S$, if $p \in T(A)$, then $p \in A$ also.

   b) S is a *locally connected* continuum if and only if $T(A) = \overline{A}$ for all $A \subseteq S$, or equivalently if $T(A) = A$ for all closed $A \subseteq S$.

   c) S is *semilocally connected at* p if and only if $T(p) = \{p\}$.

   d) S is *aposyndetic*, globally, if and only if $T(p) = \{p\}$ for all $p \in S$. (This one has a local version, too.)

   e) S is *almost connected im Kleinen* at p if and only if for each $A \subseteq S$ such that $p \in \text{Int } T(A)$, it is also true that $p \in \overline{A}$.

   f) S is *indecomposable* if and only if $T(p) = S$ for all $p \in S$.

   This list could continue somewhat longer, but these suffice to illustrate the point and are the only ones we shall use here. Part of the importance of the set function T stems from the following theorem, due to H. S. Davis, which has never appeared in print in its entirety before.

THEOREM 1: Let S and Z be continua and let f: S → Z be a continuous surjection. Then, for all $A \subseteq S$ and $B \subseteq Z$,

   a) $T(B) \subseteq fTf^{-1}(B)$.
   b) If f is monotone, $fT(A) \subseteq Tf(A)$ and $Tf^{-1}(B) \subseteq f^{-1}T(B)$.
   c) If f is monotone, $T(B) = fTf^{-1}(B)$.
   d) If f is open, $f^{-1}T(B) \subseteq Tf^{-1}(B)$.

e) If f is both monotone and open, $f^{-1}T(B) = Tf^{-1}(B)$.

Proof of d: Suppose $x \in S$ but $x \notin Tf^{-1}(B)$. Then, there is a continuum W such that $x \in Int(W)$ and $W \cap f^{-1}(B) = \phi$. Thus, $f(W) \cap B = \phi$, and since $f(x) \in f(Int(W)) \subseteq f(W)$, it follows that $f(x) \notin T(B)$. Then, $f^{-1}(f(x)) \cap f^{-1}T(B) = \phi$, so that in particular $x \notin f^{-1}T(B)$. Contraposition then yields the result.

The proofs of the other parts are similar and are left to the reader; a proof of a) can be found in [1].

As illustrations of the uses of the above theorem, consider the following well-known results.

PROPOSITION 1: Suppose S and Z are continua and $f: S \to Z$ is continuous and onto. If S is locally connected, then Z is locally connected also.

Proof: Let $B \subseteq Z$ be closed. To prove $T_Z(B) = B$ it is enough to prove that $T_Z(B) \subseteq B$. By a) of Theorem 1, $T_Z(B) \subseteq fT_S f^{-1}(B)$. By local connectedness of S, $T_S f^{-1}(B) = f^{-1}(B)$. Thus $T_Z(B) \subseteq ff^{-1}(B)$, or $T_Z(B) \subseteq B$.

PROPOSITION 2: Every monotone image of an indecomposable continuum is indecomposable.

Proof: Let $f: S \to Z$ be monotone and onto, where S is indecomposable. Suppose $p \in Z$. Then, $Tf^{-1}(p) \subseteq f^{-1}T(p)$, that is $S \subseteq f^{-1}T(p)$, and $f(S) \subseteq T(p)$. Thus $T(p) = Z$, as required.

Another, somewhat deeper result easily proven this way is:

PROPOSITION 3: Suppose $f: S \to Z$ is a continuous surjection. Suppose $K = \{p \in S | S$ is not connected im Kleinen at $p\}$ and $L = \{p \in Z | Z$ is not connected im Kleinen at $p\}$. Then $L \subseteq f(K)$.

Proof: Let $x \in L$. Then there is a closed set A such that $x \in T_Z(A) - A$. Thus since points not in K cannot lie in $T_S f^{-1}(A)$ unless they also belong to $f^{-1}(A)$, by observation a) above and our "definition" of connectedness im Kleinen, it follows that

$$x \in T_Z(A) \subseteq fTf^{-1}(A)$$
$$\subseteq f(K \cup f^{-1}(A))$$
$$\subseteq f(K) \cup ff^{-1}(A)$$
$$\subseteq f(K) \cup A .$$

But since $x \notin A$, it follows that $x \in f(K)$.

A number of other results concerning T will be listed to give an overview of the possible uses of the set function. These are only a sampling and are in no way exhaustive.

THEOREM 2: [2] If X is a continuum and A and B are closed subsets of X such that $T(A) \cap B = T(B) \cap A = \phi$ but $T(A) \cap T(B) \neq \phi$, then X is not contractible.

A well-known example illustrating this theorem is the union of two copies of the cone over the Cantor set, with only one point of their bases identified. If u and v are the vertices of the two cones, then $T(u) \cap T(v) \neq 0$, but $u \notin T(v)$ and $v \notin T(u)$, so that the continuum is not contractible.

DEFINITION: A continuum S is T-*additive* if and only if for every pair of closed sets A, B ⊆ S, we have T(A ∪ B) = T(A) ∪ T(B). S is T-*symmetric* if and only if for every pair of closed sets A and B, if A ∩ T(B) = φ, then B ∩ T(A) = φ, also.

The next theorem, due to H. S. Davis, has not appeared in print before.

THEOREM 3: If S is T-symmetric (respectively, T-additive) and f: S → Z is a continuous monotone surjection, then Z is also T-symmetric (respectively T-additive).

Proof: Suppose S is T-additive, and let A, B ⊆ Z. Applying Theorem 1, c), yields

$$T(A \cup B) = fTf^{-1}(A \cup B)$$
$$= fT(f^{-1}(A) \cup f^{-1}(B))$$
$$= f(T(f^{-1}(A)) \cup T(f^{-1}(B)))$$
$$= fTf^{-1}(A) \cup fTf^{-1}(B) = T(A) \cup T(B) .$$

The proof for T-symmetry is much the same.

As an application of this theorem, observe that since the $\sin \frac{1}{x}$ curve is T-symmetric while the harmonic fan (= the cone over ω + 1) is not, there is no monotone map from the $\sin \frac{1}{x}$ curve onto the harmonic fan. Using Theorem B of [3], it follows that a monotone image of a T-additive continuum is aposyndetic if and only if it is locally connected, a result of E. L. Vandenboss. (Note that T-additivity is defined differently in [3]; Theorem B there merely gives the equivalence of the two definitions.)

THEOREM 4: [3], [11] If $W \subseteq S$ is connected, so is $T(W)$.

Observe that T may be regarded as a function from the hyperspace of closed subsets of a continuum to itself. With this interpretation, T is always upper semicontinuous. Furthermore, if T is in fact continuous, S is either locally connected or is neither aposyndetic nor connected im Kleinen.

The next theorem is mine and is the only new result to be included in this section. Its proof is fairly representative.

THEOREM 5: If S and Z are continua, Z is locally connected, and $f: S \to Z$ is a monotone open map such that for each proper subcontinuum $W \subseteq S$, $f(W) \neq Z$, then T is continuous for S.

Proof: Suppose A is closed in S. Then since $T(f(A)) = f(A)$ by local connectedness of Z, it follows that $fTf^{-1}(f(A)) = f(A)$ by Theorem 1, c). Thus,

$$f^{-1}fTf^{-1}f(A) = f^{-1}f(A) , \quad \text{and}$$

$$T(f^{-1}f(A)) \subseteq f^{-1}f(A) .$$

Then, $f^{-1}f(A) = T(f^{-1}f(A))$ since the reverse inclusion always holds. Since $A \subseteq f^{-1}f(A)$, $T(A) \subseteq f^{-1}f(A)$.

Now, suppose there is an $x \in f^{-1}f(A)$ with $x \notin T(A)$. Let W be a continuum neighborhood of x in S missing A. Since f is open $f(x) \in \text{Int}(f(W))$, so that there is a finite collection of continua $\{M_i\}_{i=1}^{n}$ in Z such that $f(x) \notin \bigcup_{i=1}^{n} M_i$ but $Z - f(\text{Int}(W)) \subseteq \bigcup_{i=1}^{n} M_i$. This follows from the fact that $T(f(x)) = \{f(x)\}$ and a compactness argument. Furthermore, by choosing n as small as possible, we may also assume that each M_i meets $f(W)$. Then, $f(W) \cup (\bigcup_{i=1}^{n} M_i) = Z$,

and $f^{-1}f(W) \cup (\bigcup_{i=1}^{n} f^{-1}(M_i)) = S$. For each i, there is a
$q_i \in f(W) \cap M_i$, and $q_i = f(p_i)$ for some $p_i \in W$. Since
$p_i \in f^{-1}(M_i)$, $f^{-1}(M_i) \cap W \neq \phi$. Let $X = W \cup (\bigcup_{i=1}^{n} f^{-1}(M_i))$.
Then X is a subcontinuum of S. To prove $X \neq S$, recall
that $x \in f^{-1}f(A)$. Thus there is a $y \in A$ with $f(x) = f(y)$.
Then $y \notin W$ and $f(y) \notin \bigcup_{i=1}^{n} M_i$, so that $y \notin \bigcup_{i=1}^{n} f^{-1}(M_i)$ and
hence $y \notin X$. However, $f(X) = Z$, contrary to hypothesis.
Therefore, $T(A) = f^{-1}f(A)$, and $T = f^{-1} \circ f$, considered as
a hyperspace map. Since for f continuous and open each of
f^{-1} and f is continuous as a hyperspace map, the result
follows.

Remark 1: I do not know whether the condition that Z be
locally connected can be relaxed to require merely that T be
continuous for Z, nor whether every continuum S for which
T is continuous admits such a map onto a locally connected
continuum.

Remark 2: It follows from this result that the circle of
pseudo arcs [6] is a continuum for which T is continuous.

Before turning to some other set functions, I would like to
mention that in [9] H. S. Davis has precisely elucidated the
relationship between the set function T and inverse limits.
This is of major importance because it paves the way for much
deeper investigation of continua in the aforementioned "intermediate range." I have great hopes for the future of this
endeavor.

Let S be any continuum and $A \subseteq S$. Some related set functions which either have been or may be useful are defined as
follows:

K(A) = ∩{W|W is a continuum neighborhood of A}
[15], [16], [17].

Y(A) = S - {p| there is a connected open set U such that
p ∈ U and $\overline{U}$ ∩ A = φ} [18]

aT(A) = S - {p|p has a finite collection of continuum
neighborhoods, the intersection of which misses A}
[15] and [1].

aK(A) = ∩{W|A ⊆ Int(W), W is closed, and W has only
finitely many components} [15]

Γ(A) = S - {p|p has a closed neighborhood with at most
countably many components which miss A}.

All these functions except Γ have been investigated to
some extent. Although I have not studied it seriously, I sus-
pect that Γ may be useful at a slightly higher level of
pathology than T, at least for metric continua. In particular, it seems likely that Theorem 1 holds for Γ as well as
T. A brief sample of typical results involving these functions will close this section.

THEOREM 6: [17] If S is hereditarily decomposable and
p ∈ S, then Int(K(p)) = φ.

THEOREM 7: [1] S is T-additive if and only if T = aT
on S. (aT stands for "additive-T"; aK for "additive-K.")

THEOREM 8: [15] S is T-symmetric if and only if T = aK
on S.

THEOREM 9: Γ(A) ⊆ T(A) ⊆ Y(A) for all A.

Proof: This is clear. See [19] for a discussion of Y.

Example 4: None of the inclusions in Theorem 8 can be re-
versed, however. Let L_1 and L_2 be harmonic fans (cones
over simple convergent sequences with limit). Let v_1 and
v_2 denote their respective vertices and let b_1 and b_2 be
the endpoints of the limit segments in each case. Form S by
identifying b_1 and v_2.

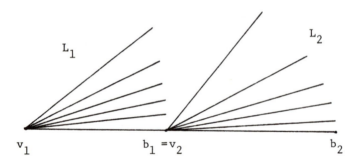

L_1

L_2

v_1 $b_1 = v_2$ b_2

Fig. 4.

Then, in S, $\Gamma(v_1) = v_1$; $T(v_1)$ is the arc $[v_1, b_1]$, while
$Y(v_1)$ is the arc $[v_1, b_2]$.

DEFINITION: A continuum S is *point T-symmetric* if and
only if for p, q ∈ S, if p ∈ T(q) then q ∈ T(p), or
equivalently if and only if K(p) = T(p) for all p ∈ S.

THEOREM 10: [1] If T is continuous on the hyperspace of
closed subsets of S, so is aT. If in addition S is point
T-symmetric, then T = aT.

It is not known whether continuity of T by itself implies that T = aT. A number of other unsolved problems pertaining to set functions are given in the problem section.

The following bibliography is not exhaustive, but it lists the articles which give the basic known properties of the set functions, as well as a few where they are used as tools for attacking other problems.

BIBLIOGRAPHY

1. Bellamy, D. P., "Continua for which the set function T is continuous." Trans. Amer. Math. Soc. 151 (1970), 581-588.

2. ------ and Charatonik, J. J., "The set function T and contractibility of continua." Bull. Acad. Polon. Sci. sér. Sci. Math. Astronom. Phys. 25 (1977), 47-49.

3. ------ and Davis, H. S., "Continuum neighborhoods and filterbases." Proc. Amer. Math. Soc. 27 (1971), 371-374.

4. ------ and Hagopian, C. L., "Mapping continua onto their cones." Colloq. Math. 41 (1979), 53-56.

5. ------ and Rubin, L. R., "Indecomposable continua in Stone-Čech compactifications." Proc. Amer. Math. Soc. 39 (1973), 427-432.

6. Bing, R. H. and Jones, F. B., "Another homogeneous plane continuum." Trans. Amer. Math. Soc. 90 (1959), 171-192.

7. Bennett, D. E., "A characterization of local connectedness by means of the set function T." Fund. Math. 86 (1974), 137-141.

8. Davis, H. S., "A note on connectedness im Kleinen." Proc. Amer. Math. Soc. 19 (1968), 1237-1241.

9. ------, "Relationships between continuum neighborhoods in inverse limit spaces and separations in inverse limit sequences." Proc. Amer. Math. Soc. 64 (1977), 149-153.

10. ------ and Doyle, P. H., "Invertible continua." Portugal Math. 26 (1967), 487-491.

11. ------, Stadtlander, D. P., and Swingle, P. M., "Properties of the set functions T^n." Portugal Math. 21 (1962), 113-133.

12. ------, Stadtlander, D. P., and Swingle, P. M., "Semi-groups, continua, and the set functions T^n." Duke Math. J. 29 (1962), 265-280.

13.   ------ and Swingle, P. M., "Extended topologies and
      iteration and recursion of set functions." Portugal.
      Math. 23 (1964), 103-129.

14.   Hunter, R. P., "On the semigroup structure of continua."
      Trans. Amer. Math. Soc. 93 (1959), 356-368.

15.   Moreland, William T., Jr., "Some properties of four set
      valued set functions." Masters' Thesis. University of
      Delaware, Newark, Delaware 19711 (1970).

16.   Rosasco, J., "A note on Jones' function K." Proc. Amer.
      Math. Soc. 49 (1975), 501-504.

17.   Schlais, H. E., "Non aposyndesis and non-hereditary de-
      composability." Pacific J. Math. 45 (1973), 643-652.

18.   Vandenboss, Eugene L., "Set functions and local connec-
      tivity." Ph.D. Dissertation, Michigan State University
      (1970), University Microfilms #71-11997.

The remaining sections are much more brief than Section I.

II. THE FIXED POINT PROPERTY

The study of fixed points of mappings is a central theme in a wide variety of mathematics. The question of which continua have the fixed point property, for either homeomorphisms or continuous self maps, has been the subject of much research for decades. The most famous unsolved problem is, of course, whether nonseparating plane continua must have the fixed point property -- problem number 107 in the Scottish Book. An excellent survey article on the fixed point property by R. H. Bing appeared in the Monthly [5]. The purpose of this section is to summarize some of the work on the subject since Bing's article appeared.

On the problem of whether nonseparating plane continua have the fixed point property, Bing has written a brief summary [4] from which I quote here on the recent work: (Reference numbers are those for this section, not Bing's summary).

> Bell [2] in 1967 and Sieklucki [13] in 1968 proved that every nonseparating plane continuum that has a hereditarily decomposable boundary has the fixed point property. Hagopian [8] in 1971 proved that every arcwise connected nonseparating plane continuum has the fixed point property. In 1972 Hagopian [7] extended his theorem to every nonseparating plane continuum with the property that every pair of its points can be joined by a hereditarily decomposable subcontinuum. [This result was also obtained by J. Krasinkiewicz [10]] . . .
> Recently Bell [1] proved that every homeomorphism of a nonseparating plane continuum onto itself that can be extended to [a homeomorphism of] the plane has a fixed point.

On the other problems, there have been at least four major developments in the past six years: L. Mohler [12] has shown that uniquely arcwise connected continua have the fixed point property for homeomorphisms. C. L. Hagopian [9] proved that uniquely arcwise connected plane continua have the fixed point property. Mohler's techniques were measure-theoretic, while Hagopian's involved very delicate geometric arguments in the plane and some inverse limit techniques. R. Mańka [11] gave a proof that hereditarily decomposable hereditarily unicoherent continua, called λ-dendroids, have the fixed point property, using a major refinement of what Bing [5] calls the "dead end" argument. H. Cook [6] has shown that these continua are necessarily tree-like.

Using simple algebraic properties of solenoids and compactification techniques, I constructed an example of a fixed point free map on a tree-like continuum [3].

The questions of which continua have the fixed point property and what sorts of maps on tree-like or other continua necessarily have fixed points are the subject of much continuing research.

BIBLIOGRAPHY

1. Bell, H., "A fixed point theorem for plane homeomorphisms."
    Bull. Amer. Math. Soc. 82 (1976), 778-780.

2. ------, "On fixed point properties of plane continua."
    Trans. Amer. Math. Soc. 128 (1967), 539-548.

3. Bellamy, D. P., "A tree-like continuum without the fixed
    point property." Houston J. Math. 6 No. 1 (1980), 1-13.

4. Bing, R. H., "Does every nonseparating plane continuum
    have the fixed point property?" (A historical synopsis
    of work on the problem.) Preprint.

5. ------, "The elusive fixed point property." Amer. Math.
    Monthly 76 (1969), 119-132.

6. Cook, H., "Tree-likeness of dendroids and λ-dendroids."
    Fund. Math. 68 (1970), 19-22.

7. Hagopian, C. L., "Another fixed point theorem for plane
    continua." Proc. Amer. Math. Soc. 31 (1972), 627-628.

8. ------, "A fixed point theorem for plane continua."
    Bull. Amer. Math. Soc. 77 (1971), 351-354.

9. ------, "Uniquely arcwise connected plane continua have
    the fixed point property." Trans. Amer. Math. Soc. 248
    (1979), 85-104.

10. Krasinkiewicz, J., "Concerning the boundaries of plane
    continua and the fixed point property." Bull. Acad.
    Polon. Sci. Sér. Sci. Math. Astronom. Phys. 21 (1973),
    427-431.

11. Mańka, R., "Association and fixed points." Fund. Math.
    91 (1976), 105-121.

12. Mohler, L., "The fixed point property for homeomorphisms
    of 1-arcwise connected continua." Proc. Amer. Math. Soc.
    52 (1975), 451-456.

13. Sieklucki, "On a class of plane acyclic continua with the
    fixed point property." Fund. Math. 63 (1968), 257-278.

III. HOMOGENEITY

A topological space X is *homogeneous* if and only if whenever
x, y ∈ X, there is a homeomorphism h: X ↠ X such that
h(x) = y. There has been an explosion of work in this area
in recent years, largely because of a result by E. G. Effros
in the theory of transformation groups [3].

THEOREM 1 (Effros): Let X be a complete separable metric
space and let G be a complete separable metric group of
homeomorphisms of X. Then the following statements are
equivalent:

a) Each orbit Gx [$Gx = \{g(x) \mid g \in G\}$] is a G_δ subset
of X.

b) The orbit space, X/G, is a T_0 space.

c) For each x ∈ X the evaluation map $\alpha_x: G \to Gx$ defined by $\alpha_x(h) = h(x)$ is an open mapping.

C. L. Hagopian [4] Lemma 4, p. 37, proved a corollary to
this theorem, from which the following result follows easily.

THEOREM 2 (Hagopian): If (X,d) is a homogeneous compact
metric space, then given any ε > 0, there is a δ > 0 such
that if x, y ∈ X and d(x,y) < δ, then there is a homeomorphism h: X → X, within ε of the identity, such that
h(x) = y.

Indication of proof: The proof of this remarkable result
is fairly easy from Effros' theorem, since the group $\mathcal{H}$ of
all homeomorphisms of X is always complete and separable and
for x ∈ X, $\mathcal{H}x = X$ which is certainly a G_δ. The result
then follows from c) of Theorem 1, together with a standard

Lebesgue number argument to yield a single δ for all points in X.

Theorem 2 has proven most useful in the study of homogeneity, although the stronger Theorem 1 seems to be needed in some cases. The first application of this result to the study of homogeneity appears in [6], where G. S. Ungar proves, among other things, that a two–homogeneous continuum is locally connected, settling a long outstanding question. I shall give a proof of a slightly weaker result here as an indication of the methods.

DEFINITION: A topological space X is *two-homogeneous* if and only if given any a, b, u, v $\in$ X there is a homeomorphism h: X $\to$ X such that h({a,b}) = {u,v}. X is *strongly* two-homogeneous if and only if h can always be chosen so that h(a) = u and h(b) = v.

THEOREM 3: (Ungar -- actually Ungar proved more.) Every strongly two-homogeneous metric continuum is locally connected.

Proof: Suppose x $\in$ X and U is open in X. Let V be open such that x $\in$ V $\subseteq$ $\overline{V}$ $\subseteq$ U. Let $\mathcal{H}$ denote the group of all homeomorphisms of X and let G = {h $\in$ $\mathcal{H}$|h(x) = x}. G is closed in $\mathcal{H}$ and so is complete. By strong two-homogeneity, the orbits under G are {x} and X - {x}, both G_δ's, so that for any y $\in$ X - {x}, the map α: G $\to$ X - {x} defined by α(h) = h(y) is open. In particular, let y $\neq$ x be a point in the component, K, of $\overline{V}$ containing x, and define

$$\mathcal{O} = \{h \in G | h(\overline{V}) \subseteq U\} \ .$$

Then $\mathcal{O}$ is open in G, so that $\alpha(\mathcal{O})$ is open in X - {x} and hence in X. But, $\alpha(\mathcal{O}) \subseteq$ U and if p $\in$ $\alpha(\mathcal{O})$, there is

an h ∈ G with h(y) = α(h) = p, and h(K) is a connected
subset of U joining x and p. Thus the component of U
containing x has nonempty interior. It then follows from a
standard "closing down" argument that X is connected im
Kleinen at some point (see [2]) and by homogeneity at every
point, which yields that X is locally connected, completing
the proof.

If it can be shown that the only homogeneous nonseparating
plane continuum is the pseudo-arc, the characterization of
plane homogeneous continua will be complete. This subject is
discussed elsewhere in this volume by I. W. Lewis. As the
final draft of this manuscript was in preparation, I received
a preprint [5] in which it is shown that if X is a homogeneous plane continuum which does not separate the plane,
then the span of X is zero. This means that if S is any
continuum and f, g: S → X are continuous functions with the
same image in X, then for some x ∈ S, f(x) = g(x).

There is also some reason to hope for the eventual classification of all one-dimensional homogeneous metric continua.
This is discussed in the problem section.

For continua not assumed to be either planar or one-dimensional, homogeneity is much less well understood. I believe
that one of the most interesting and important problems here
is whether arcwise connected homogeneous metric continua are
necessarily locally connected. L. Lum and I [1] have shown
that each two points of such a continuum lie together on a
simple closed curve. Many deep and interesting problems remain in this area.

BIBLIOGRAPHY

1. Bellamy, D. P., and Lum, L., "The cyclic connectivity of
    arcwise connected homogeneous continua." Trans. Amer.
    Math. Soc. (to appear).

2. Davis, H. S., and Doyle, P. H., "Invertible continua."
    Portugal. Math. 26 (1967), 487-491.

3. Effros, E. G., "Transformation groups and C* algebras."
    Ann. of Math. 81 (1965), 38-55.

4. Hagopian, C. L., "Homogeneous plane continua." Houston
    J. Math. 1 (1975), 35-41.

5. Oversteegen, L. G., and Tymchatyn, E. D., "Plane strips
    and the span of continua." (Preprint)

6. Ungar, G. S., "On all kinds of homogeneous spaces." Trans.
    Amer. Math. Soc. 212 (1975), 393-400.

IV. CONTINUOUS MAPPINGS

Here I want merely to list a few observations and give a proof of one result which further illustrates the methods of Section I.

Let X and Y be continua. Under what conditions is there a continuous map of X onto Y? The attempt to answer this question forms a major part of modern continua theory. Here I am going to informally list some observations which are useful in this regard. Other aspects of the question are discussed in the related paper [1] and in the excellent monograph [4].

Assume that X and Y are continua and that Y is a continuous image of X. It is a standard result that if X is arcwise connected so is Y. Further, if X has a dense arc component, so does Y, and if every arc component of X is dense, then every arc component of Y is dense, also. The number of arc components of Y cannot exceed the number possessed by X. However, the number of dense arc components can be either increased or decreased.

Example: Let $A \cup B \cup C = X$ where each of A, B, and C is a copy of the dyadic solenoid, where $C \cap A$ and $C \cap B$ are each singletons, and lie in different composants of C. Then X has no dense arc components, but th⌐ number can be increased to any finite number or a denumerable infinity merely by shrinking C to a point and then identifying as many pairs of points a_j and b_j as desired with $a_j \in A$, $b_j \in B$, such that no two of the a_j's (b_j's) lie in the same composant of A (B). (Of course, if there are infinitely many of the a_j's and b_j's, the sets must be closed and must be chosen homeomorphic under $a_j \to b_j$.) To obtain uncountably many dense arc components is even easier: Simply begin with X

and shrink B ∪ C to a point. To decrease the number of
dense arc components, if a continuum has n of them select a
point from each of k of these n and identify these k
points to a single point, obtaining a quotient continuum with
n − k + 1 dense arc components.

DEFINITION: A continuum is *weakly chainable* if and only if
it is a continuous image of the pseudo arc [2], [3]. (This is
not the original definition, but a characterization due to L.
Fearnley and A. Lelek. However, it is most useful for us.)

It can be shown that a continuum which is a finite union of
weakly chainable continua is weakly chainable. There is an
exact analog to arcwise-connectedness arising from the notion
of weak chainability:

DEFINITION: A topological space is F − L connected (for
Fearnley-Lelek) if and only if each pair of points in X lies
in a weakly chainable subcontinuum of X.

One can then define F − L components of a continuum in the
same manner as arc components are defined. All the above observations concerning arcwise-connectedness then can be re-
stated for F − L connectedness. F − L connected continua bear
the same relationship to weakly chainable continua that arcwise connected continua bear to Peano Continua:

THEOREM 1: A continuum X is F − L connected (arcwise connected) if and only if each two points of X lie together in
a weakly chainable (Peano) subcontinuum of X.

Note that in the example, above, the F − L components and
arc components are the same, though in general F − L components
may be larger.

The following final result seems likely to be known, although I do not know of a reference. I am including it as an illustration of the techniques of Section I and as an introduction to the conjecture which will close this article.

THEOREM 2: Let X and Y be countable compact Hausdorff spaces. If the derived set X' is finite while Y' is infinite, then the cone over X cannot be mapped onto the cone over Y.

Proof: $C(X')$ is the closure of the set of points at which $C(X)$ is not connected im Kleinen; similarly for $C(Y')$. Thus, if $f: C(X) \to C(Y)$ were a continuous surjection, we would have by Proposition 3 of Section I and continuity,

$$C(Y') \subseteq f(C(X')) .$$

However $f(C(X'))$ is a Peano continuum, while $C(Y')$ is not contained in any locally connected subset of $C(Y)$, a contradiction.

As a generalization of this, it seems likely that the following can be proven by transfinite induction.

Conjecture: If X and Y are compact Hausdorff scattered spaces and α is an ordinal such that the α-*th* derived set of X is finite while the α-*th* derived set of Y is infinite, then there is no mapping of $C(X)$ onto $C(Y)$.

In the metric case, such spaces X and Y are homeomorphic to countable non-limit ordinals, so this would be a first step toward understanding the continuous images and preimages of the cones over these spaces. Little is known about this subject. It seems a very natural place to work in an attempt to understand continua of intermediate complexity.

BIBLIOGRAPHY

1. Bellamy, D. P., "Continuous mappings between continua."
    Proceedings: Conference on Metric Spaces, Generalized
    Metric Spaces, and Continua. Department of Math, Guilford
    College, Greensboro, NC 27410. Compiled by G. R. Gordh,
    Jr. and J. R. Boyd. (To appear)

2. Fearnley, L., "Characterizations of the continuous images
    of the pseudo-arc." Trans. Amer. Math. Soc. 111 (1964),
    380-399.

3. Lelek, A., "On weakly chainable continua." Fund. Math.
    51 (1962-63), 271-282.

4. Maćkowiak, T., "Continuous mappings on continua." Disser-
    tationes Mathematicae 158. PWN (Polish Scientific Pub-
    lishers) Warszawa (1979).

Stable Homeomorphisms, Galois Spaces, and Related Properties in Homogeneous Continua

Wayne Lewis and Judy Kennedy Phelps

Fletcher and Snider introduced the study of Galois spaces in [6] and Fletcher introduced the study of representable spaces in [5]. Both concepts involve the existence of certain types of stable homeomorphisms. We are interested in their role in homogeneous continua, since they seem closely related to information about the structure and types of homogeneity in such continua. This paper contains some observations about the existence and differences between these properties in homogeneous continua, and raises several questions.

A homeomorphism $h : M \to M$ is *primitively stable* if there exists a non-empty open set U in M such that $h|_U = id_U$. A *stable* homeomorphism is a finite composition of primitively stable homeomorphisms. A space X is *Galois* if for each $x \in X$ and open U containing x there exists a homeomorphism $h : X \to X$ with $h(x) \neq x$ and $h|_{X-U} = id_{X-U}$. A space X is *representable* if for each $x \in X$ and open U containing x there exists an open V, with $x \in V \subset U$, such that for each $y \in V$ there is a homeomorphism $h : X \to X$ with $h(x) = y$ and $h|_{X-U} = id_{X-U}$. (Representability is equivalent to strong local homogeneity [7]. This is shown in [2].)

Consider the following three statements about a homogeneous continuum N.

1.) N admits a non-identity stable homeomorphism.

2.) N is a Galois space.

3.) N is representable.

Clearly 3.) $\to$ 2.) $\to$ 1.). In [11] it was shown that for a 2-homogeneous continuum these statements are all equivalent.

If this is weakened to considering locally connected homogeneous continua, then none of these statements are equivalent, as the following examples show.

Example A. The product of two or more copies of the Menger universal curve admits a non-identity stable homeomorphism but is not Galois.

If $h : M^\alpha \to M^\alpha$ is a homeomorphism of $\prod_{n<\alpha} M_n$ with each M_n a copy of the Menger curve and $2 \leq \alpha \leq \omega_0$, then $h = so(\prod_{n<\alpha} h_n)$ where s is a permutation of the coordinates of $\prod_{n<\alpha} M_n$ and each h_n is a homeomorphism of M_n [8],[10]. Let $\{x_i\}_{i<\alpha} \in M^\alpha$ and $U = \prod_{n<\alpha} U_n$ be an open set containing $\{x_i\}_{i<\alpha}$ with $U_0 \neq M_0$ and $U_1 \neq M_1$. Let h be a homeomorphism of M^α which is the identity outside of U. There exist points $\{y_i\}_{i<\alpha}$ and $\{z_i\}_{i<\alpha}$, both outside of U, so that $y_i = z_i$ if and only if $i = 0$. Since $h(\{y_i\}_{i<\alpha}) = \{y_i\}_{i<\alpha}$ and $h(\{z_i\}_{i<\alpha}) = \{z_i\}_{i<\alpha}$, it follows that $s(0) = 0$. Also for each $n < \alpha$ there are points $\{a_i\}_{i<\alpha}$ and $\{b_i\}_{i<\alpha}$, both outside of U, with $a_i = b_i$ if and only if $i = 0$ or $i = n$. Thus $s(n) = n$ and s is the identity permutation. For each $n < \alpha$ there exists a point $\{c_i\}_{i<\alpha}$ not in U with $c_n = x_n$. (Note that we need $c_1 \notin U_1$ if $n = 0$ and $c_0 \notin U_0$ if $n = 1$.) Since $h(\{c_i\}_{i<\alpha}) = \{c_i\}_{i<\alpha}$, and s is the identity, we have $h_n(x_n) = x_n$. Thus h is the identity, and M^α is not Galois. Since M is representable [1], M^α does admit non-identity stable homeomorphisms. □

This example provides a negative answer to the question by Duvall, Fletcher, and McCoy [4] of whether every homogeneous continuum is Galois. After the authors produced Example A it was noted that the case for $\alpha = 2$ had been observed by W.

Kuperberg at a war eagle at Auburn a few years ago. The
observation has not appeared in print.

Example B. The product $M \times S^1$ of the Menger curve and
the simple closed curve is Galois but is not representable.

Let $(x,y) \in M \times S^1$ and $U \times V$ be a basic open set containing (x,y). Let $F : S^1 \times [0,1] \to S^1$ be an isotopy with
$F_0 = id_{S^1}$, $F_t(Z) = Z$ for all $Z \notin V$ and all $t \in [0,1]$, and
$F_1(y) \neq y$. Let $g : M \to [0,1]$ be a map with $g(w) = 0$ for
all $w \notin U$ and $g(x) = 1$. Define $h : M \times S^1 \to M \times S^1$ by
$h((a,b)) = (a,F_{g(a)}(b))$. Then $h((x,y)) \neq (x,y)$ and
$h((c,d)) = (c,d)$ for all $(c,d) \notin U \times V$. Thus $M \times S^1$ is a
Galois space [4]. The Kuperbergs and Transue [8] have shown
that $M \times S^1$ is not 2-homogeneous (in fact each homeomorphism
of it preserves S^1 fibers), and thus not representable. □

Definition. [4] $\overline{X}$ is an _isotopy Galois space_ if for
each $x \in \overline{X}$ and U open containing x there exists an
isotopy $F : \overline{X} \times [0,1] \to \overline{X}$ with $F_0 = id_{\overline{X}}$, $F_1(x) \neq x$, and
$F_t|_{\overline{X}-U} = id_{\overline{X}-U}$ for each $t \in [0,1]$.
In Example B we used the fact that S^1 is an isotopy
Galois space to show that $M \times S^1$ is an isotopy Galois space.
S^1 could have been replaced by any other isotopy Galois continuum $\overline{X}$ to give an equivalent result. (It was shown in
[11] that $M \times \overline{X}$ is never representable if $\overline{X}$ is a non-
degenerate continuum.)

Question 1. Is there a (homogeneous) continuum N such
that $M \times N$ is Galois but not isotopy Galois?

Duvall, Fletcher, and McCoy [4] have presented an example
of a connected, metric Galois space which is not an isotopy
Galois space. They asked whether there exists a Galois

continuum which is not an isotopy Galois space, and suggested the pseudo-arc as a possible candidate since it clearly admits no isotopies. In [9] it is shown that the pseudo-arc is Galois, providing a positive answer to this question. It follows from [3] and [8] that the Menger curve admits no isotopies, and so is an example of a representable continuum which is not an isotopy Galois space.

Implicit in some earlier work on locally connected homogeneous continua is the following question.

Question 2. When is a locally connected homogeneous continuum 2-homogeneous?

At present the only known non-examples are of the form $M \times \overline{X}$, where M is the Menger curve and $\overline{X}$ is a non-degenerate locally connected homogeneous continuum. (Such a product is never 2-homogeneous.)

In the stable homeomorphism result on the pseudo-arc, it was actually shown that the pseudo-arc satisfies a stronger condition than being Galois.

Condition *: $\overline{X}$ is a metric space such that for any subcontinuum N of $\overline{X}$, points p, q of N, and $\varepsilon > 0$, there exists a homeomorphism $h : \overline{X} \to \overline{X}$ with $h(p) = q$ and $h(z) = z$ for each z outside the ε-neighborhood of N.

For continua, satisfying condition * implies being a Galois space. All known 2-homogeneous continua or indecomposable homogeneous continua satisfy condition *. ($M \times S^1$ does not.)

Observation. If $\overline{X}$ is a locally connected, complete metric space then $\overline{X}$ satisfies condition * if and only if $\overline{X}$ is representable.

(Clearly being representable implies condition *. Locally connected, complete metric spaces are locally arcwise connected so, by choosing appropriate small neighborhoods and arcs lying in them, condition * implies representability.)

As noted above, all known homogeneous indecomposable continua are Galois and satisfy condition *. (The solenoids are in fact isotopy Galois.)

Question 3. Is every homogeneous (hereditarily) indecomposable continuum Galois? Does it satisfy condition *?

Every known homogeneous continuum admits a non-identity stable homeomorphism, leading to the following questions.

Question 4. [11] Does every 2-homogeneous continuum admit a non-identity stable homeomorphism?

Question 5. Does every homogeneous continuum admit a non-identity stable homeomorphism?

An affirmative answer to question 4 would show that 2-homogeneity is equivalent to representability in continua [11], (implying n-homogeneity for all n, countable dense homogeneity, and other nice properties), and that there is a sharp difference between 2-homogeneous continua and locally connected homogeneous continua.

The following questions are also of interest.

Question 6. If N is a homogeneous continuum admitting a non-identity stable homeomorphism h supported on the open set U, x ∈ U, h(x) ≠ x, then is the set of all points y in U for which there exists a stable homeomorphism g_y supported on U with $g_y(x) = y$ dense in itself? (Yes, if N is Galois.) Is it connected?

Question 7. If N is a homogeneous continuum such that
N × N admits a non-identity stable homeomorphism does N
admit a non-identity stable homeomorphism?

A special case of question 7 would be whether homeotopi-
cally homogeneous continua admit non-identity stable homeomorphisms.

Question 8. If N × N is 2-homogeneous, is the continuum
N isotopically representable?

Information about how the set of all stable homeomorphisms
of a homogeneous continuum N lies in the space of all homeomorphisms of N, or in fact any other information about the
space of homeomorphisms of N, would be desirable.

R. D. Anderson [3] has shown that the space of homeomorphisms of the Menger curve is totally disconnected. For
orientable manifolds all stable homeomorphisms lie in the
same component of the space of homeomorphisms as the identity.

In [9] the following question is asked, for which it is
conjectured that the answer is no. There is, of course, no
need to restrict the question to the pseudo-arc.

Question 9. [9] Is the set of stable homeomorphisms dense
in the space of all homeomorphisms of the pseudo-arc?

At present very little is known about stable homeomorphisms
for homogeneous continua in the range between indecomposability and local connectivity.

Looking at almost homogeneous continua, the result on
stable homeomorphisms of the pseudo-arc [9] also shows that
any continuum every non-degenerate proper subcontinuum of
which is a pseudo-arc satisfies condition *. At the other
end of the spectrum, the Sierpinski universal curve is almost
homogeneous and Galois [3], while its product with itself

admits a non-identity stable homeomorphism but is not Galois
[10]. The space of homeomorphisms of the Sierpinski curve
is also totally disconnected [3].

REFERENCES

[1] Anderson, R. D., "1-dimensional continuous curves and a
     homogeneity theorem." Ann. Math. 68 (1958), 1-16.

[2] Bales, John W., "Representable and strongly locally
     homogeneous spaces and strongly n-homogeneous spaces."
     Houston J. Math. 2 (1976), 315-327.

[3] Brechner, Beverly L., "On the dimensions of certain
     spaces of homeomorphisms." Trans. Amer. Math. Soc. 121
     (1966), 516-548.

[4] Duvall, Paul F., Peter Fletcher, and Robert A. McCoy,
     "Isotopy Galois spaces." Pacific J. Math. 45 (1973),
     435-442.

[5] Fletcher, Peter, "Notes on quasi-uniform spaces and
     representable spaces." Colloq. Math. 23 (1971), 263-
     265.

[6] Fletcher, P. and R. L. Snider, "Topological Galois
     spaces." Fund. Math. 68 (1970), 143-148.

[7] Ford, L. R., Jr., "Homeomorphism groups and coset
     spaces." Trans. Amer. Math. Soc. 77 (1954), 490-497.

[8] Kuperberg, K., W. Kuperberg, and W. R. R. Transue, "On
     the 2-homogeneity of Cartesian products." Fund. Math.,
     (to appear).

[9] Lewis, Wayne, "Stable homeomorphisms of the pseudo-arc."
     Can. J. Math. 31 (1979), 363-374.

[10] Phelps, Judy Kennedy, "Homeomorphisms of products of
     universal curves." Houston J. Math. 6 (1980), 127-134.

[11] ------, "A condition under which 2-homogeneity and rep-
     resentability are the same in continua." Fund. Math.,
     (to appear).

Principal Embeddings of Atriodic Plane Continua

John C. Mayer

1. INTRODUCTION

Known examples of tree-like continua which admit fixed point
free maps are atriodic, nonchainable, and each proper subcontinuum is an arc [2,13]. In [5] Brechner and Mayer show that
if there is an indecomposable nonseparating plane continuum
which admits a fixed point free map, it must have a Lake-of-
Wada channel in every embedding. This is a consequence of independent results of Bell [1] and Sieklucki [14]. In this paper
an example is given of a tree-like, atriodic, nonchainable,
indecomposable nonseparating plane continuum each of whose
proper subcontinua is an arc which has an embedding with a
Lake-of-Wada channel. The construction of the example is based
upon an example of Ingram's [8]. Ingram's example is a tree-
like, atriodic, nonchainable, nonseparating plane continuum
each of whose proper subcontinua is an arc. However, since it
has an embedding with no Lake-of-Wada channel, it has the fixed
point property [5, Th. 4.1]. Both Ingram's example and ours
are proved nonchainable by showing they have a positive span.

We construct our example as an inverse limit of X's, so may
refer to it as the X-odic continuum. The resulting continuum,
X, is homeomorphic to a continuum defined as the intersection
of a defining sequence (in the sense of Ingram and Cook [9])
of tree-covers, and we make use of both constructions in our
proofs.

We understand C. Hagopian has a similar example as an inverse limit of X's, but with exactly two Lake-of-Wada channels.

In Section 2 we construct the example, X, and prove that it
is atriodic and nonchainable. The main theorems of this section

are 2.4 showing that X is atriodic, and 2.7 showing that X is
nonchainable. The most complex theorem is 2.6 showing that X
has properties sufficient to guarantee positive span.

In Section 3 we show that X has an embedding in the plane
with a Lake-of-Wada channel. We also raise some questions
about X and the fixed point problem for nonseparating plane
continua.

All spaces are metric and distance functions are as usual
for spaces and their products. All functions (maps) are
continuous.

2. THE X-ODIC CONTINUUM

We shall define the X-odic continuum X in two ways. It is
evident that the continua so defined are homeomorphic. The
second definition will be such that $X \subseteq E^2$, and so provides
an embedding of X in the plane.

2.1. Inverse Limit Definition of X.

Let X_1 be the union of intervals [-1,1] on the coordinate
axes in the xy-plane. For convenience we designate (0,1)   as
A, (-1,0) as B, (1,0) as C, (0,-1) as D, and (0,0) as O. X_1
is then the identification of four intervals, [OA], [OB],
[OC], [OD], at a single point O. By $\frac{pA}{q}$ we mean the point
$(0,\frac{p}{q})$ and similarly for B, C, and D. Thus $[0 \ \frac{pC}{q}]$ denotes
the interval on the x-axis from (0,0)   to $(\frac{p}{q},0)$.

Let $f : X_1 \rightarrow X_1$ be a map carrying:

$0 \ \frac{A}{6}$ onto OB, OR   (order-reversing and proportionally)

$\frac{A}{6} \ \frac{A}{3}$ onto OA, OP   (order-preserving and proportionally)

$\frac{A}{3} \ \frac{A}{2}$ onto OA, OR

$\frac{A}{2} \ \frac{2A}{3}$ onto OC, OP

$\dfrac{2A}{3}\,\dfrac{5A}{6}$ onto   OC, OR

$\dfrac{5A}{6}\,A$ onto   OD, OP

$0\,\dfrac{B}{3}$ onto   OB, OR

$\dfrac{B}{3}\,\dfrac{B}{2}$ onto $0\dfrac{A}{3}$, OP

$\dfrac{B}{2}\,\dfrac{2B}{3}$ onto $0\dfrac{A}{3}$, OR

$\dfrac{2B}{3}\,B$ onto   OC, OP

$0\,\dfrac{C}{2}$ onto   OB, OR

$\dfrac{C}{2}\,0$ onto   OD, OP

$0\,\dfrac{D}{2}$ onto   OB, OR

$\dfrac{D}{2}\,D$ onto   OC, OP

A schematic diagram of f is given in Figure 1.

For each i, let $X_i = X_1$ and $f_i = f$. We define X by:

$$X = \lim_{\leftarrow} \{X_i, f_i\}$$

Let $f_1^n : X_n \to X_1$ be defined by:

$$f_1^n = f_1 \circ f_2 \circ f_3 \circ \cdots \circ f_n$$

2.2 Defining Sequence for X.

For each n, let T_n be a collection of open disks in the plane, E^2 such that:

(1) T_{n+1} strongly refines T_n; that is, for each $L \in T_{n+1}$ there is some $M \in T_n$ such that $\overline{L} \subseteq M$.

(2) $\mu(T_n) \leq \dfrac{1}{2^n}$.

(3) T_n is a coherent collection of four subchains with exactly one junction link designated as follows:

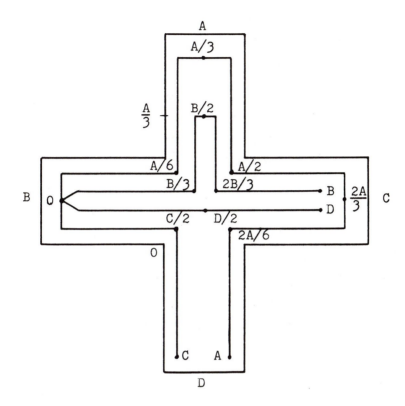

Fig. 1.

(a) The junction link is O_n.

(b) The four endlinks are A_n, B_n, C_n, and D_n.

(c) The four subchains are $O_n A_n$, $O_n B_n$, $O_n C_n$, and $O_n D_n$.

(4) T_{n+1} follows in T_n the pattern suggested by function f, including orientation with respect to the plane. (That is, if T_n were "straightened out" by some orientation-preserving homeomorphism of E^2, then T_{n+1} would sit in T_n exactly as Figure 1 suggests.)

Then $X = \bigcap\limits_{n=1}^{\infty} T_n$.

Figure 2 illustrates the first three stages in the construction of X. The third stage is represented by its nerve. Since T_{n+1} follows the pattern of f in T_n, we may conveniently refer to links of T_n on analogy with points of X_n. That is, $\frac{A_{n+1}}{3}$ is a link of chain $O_{n+1} A_{n+1}$ that corresponds to point $(0, \frac{A}{3})$ of X_n. Note $\frac{A_{n+1}}{3}$ is a link of T_{n+1} sitting in link A_n of T_n. Similarly, link $\frac{B_{n+1}}{2}$ of T_{n+1} is a subset of link $\frac{A_n}{3}$ of T_n. We could have defined the pattern that T_{n+1} follows in T_n directly in terms of such intermediate links and the chains between them. We require our chains have as *few* bends as possible. (Too many bends and we might get pseudo-arcs as subcontinua.) We will call each T_n (which is a tree cover of X) an <u>X-od</u> cover of X. An open cover of X by subsets of X is derived from the above by letting T_n' be the collection of open sets of X such that:

$$L' \in T_n' \quad \text{iff} \quad L' = L \cap X \quad \text{for some} \quad L \in T_n$$

It is clear from our definition that $X \subseteq E^2$ in the construction of 2.2. We will refer to this embedding of X in what follows.

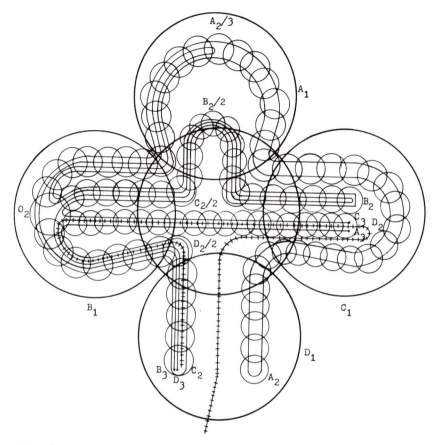

Fig. 2.

2.3. Definitions

(Ingram [8, p. 99], due to Lelek [10].) The span, σf, of
function $f : X \to Y$ is the least upper bound of all numbers ε
for which there is a connected subset Z_ε of $X \times X$ such that
$\pi_1(Z_\varepsilon) = \pi_2(Z_\varepsilon)$ and $d(f(x), f(y)) \geq \varepsilon$ for all $(x,y) \in Z_\varepsilon$.
The span, σX, of space X is the span of the identify func-
tion on X.

2.4. Theorem

Let X be the continuum defined by X-od tree covers in 2.2.
Then X is atriodic.

   Proof. We will show every proper subcontinuum of X is
chainable. That X is then atriodic follows from Ingram
[7, Th. 3].

   Let H be a proper subcontinuum of X. Let $\{T_n\}_{n=1}^{\infty}$ be
the defining sequence of tree-covers for X, defined in 2.2.
For each n, let F_n be that subcollection of T_n that covers
H minimally. Then $H = \bigcap_{n=1}^{\infty} F_n$.

   For some N, for all $k \geq N$, $F_k \neq T_n$, for otherwise
$H = M$. Since H is a continuum, each F_n must be coherent.
We consider two cases: If for some subsequence $\{F_j\}_{j=1}^{\infty}$ of
$\{F_n\}_{n=1}^{\infty}$, F_j does not include 0_j, then $H = \bigcap_{j=1}^{\infty} F_j$ is
chainable as each F_j is a chain.

   So assume there is no such subsequence. Then for some J,
for all $k \geq J$, $0_k \in F_k$. Let k be chosen as the maximum
of N, J.

   Then $0_{k+3} \in F_{k+3}$. Now $0_{k+3} \subseteq B_{k+2} \in F_{k+2}$. But
$0_{k+2} \in F_{k+2}$. Since F_{k+2} is coherent, chain $0_{k+2}B_{k+2} \subseteq F_{k+2}$.

   Now $0_{k+2}B_{k+2} \subseteq 0_{k+1}B_{k+1} \cup 0_{k+1}\dfrac{A_{k+1}}{3} \cup 0_{k+1}C_{k+1}$,

and $0_{k+1}B_{k+1} \subseteq 0_k B_k \qquad \cup 0_k \dfrac{A_k}{3} \qquad \cup 0_k C_k$, while $O_{k+1} \dfrac{A_{k+1}}{3} \subseteq O_k B_k \quad \cup O_k A_k$,

and $O_{k+1} C_{k+1} \subseteq O_k B_k \quad \cup O_k D_k$

Hence, $O_{k+2} B_{k+2}$ meets every element of T_k. Since $O_{k+2} B_{k+2} \subseteq F_{k+2}$ and F_k contains F_{k+2}, $F_k = T_k$. But $k \geq N$, so $F_k \neq T_k$, a contradiction.

Therefore, H is chainable, and by Ingram's theorem noted above, atriodic. QED

Figure 3 shows how $O_3 B_3$ sits in T_1.

To show that X is nonchainable we will use a theorem of Ingram's [8, Th. 4]:

2.5.   Theorem.

If $X = \lim_{\leftarrow} \{X_i, f_i\}$ with each X_i compact, and for $\varepsilon > 0$, $\sigma f_1^n \geq \varepsilon$, for each n, then $\sigma X > 0$.
Lelek [10, p. 210] observes that $\sigma X > 0$ implies X is nonchainable.

To apply 2.5 we require the following as a lemma. The proof is an application of the method of proof of Theorem 2 in [8].

2.6.   Theorem.

There exists a sequence Z_1, Z_2 ••• of subcontinua of $X_1 \times X_1$ such that for each n, $\pi_1(Z_n) = \pi_2(Z_n) = X_1$, $f \times f(Z_{n+1}) = Z_n$, $Z_n = Z_n^{-1}$, and if $(p,q) \in Z_1$, then $d(p,q) \geq \frac{1}{3}$. Thus $\sigma f^n \geq \frac{1}{3}$, for all n.

Proof.   Our proof is by induction on n. Z_1 is the union of the following twenty subcontinua of $X_1 \times X_1$:

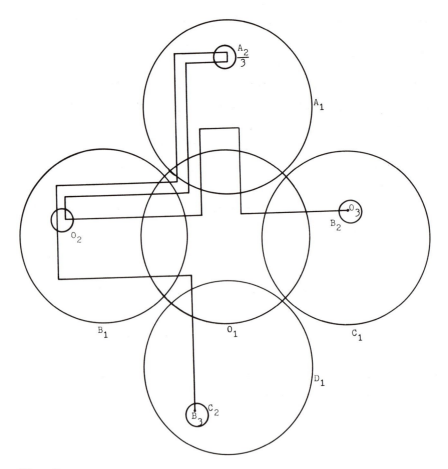

Fig. 3.

m_1 = ([OB] × {A}) ∪ ({B} × [OA])

m_2 = ([OA] × {B}) ∪ ({A} × [OB])

m_3 = ([OB] × {$\frac{A}{3}$}) ∪ ({B} × [O$\frac{A}{3}$])

m_4 = ([O$\frac{A}{3}$] × {B}) ∪ ({$\frac{A}{3}$} × [OB])

m_5 = ([OC] × {A}) ∪ ({C} × [OA])

m_6 = ([OA] × {C}) ∪ ({A} × [OC])

m_7 = ([OC] × {$\frac{A}{3}$}) ∪ ({C} × [O$\frac{A}{3}$])

m_8 = ([O$\frac{A}{3}$] × {C}) ∪ ({$\frac{A}{3}$} × [OC])

m_9 = ([OD] × {A}) ∪ ({D} × [OA])

m_{10} = ([OA] × {D}) ∪ ({A} × [OD])

m_{11} = ([OD] × {$\frac{A}{3}$}) ∪ ({D} × [O$\frac{A}{3}$])

m_{12} = ([O$\frac{A}{3}$] × {D}) ∪ ({$\frac{A}{3}$} × [OD])

m_{13} = ([OC] × {B}) ∪ ({C} × [OB])

m_{14} = ([OB] × {C}) ∪ ({B} × [OC])

m_{15} = ([OD] × {B}) ∪ ({D} × [OB])

m_{16} = ([OB] × {D}) ∪ ({B} × [OD])

m_{17} = ([OD] × {C}) ∪ ({D} × [OC])

m_{18} = ([OC] × {D}) ∪ ({C} × [OD])

m_{19} = ([O$\frac{A}{3}$] × {A}) ∪ ({D} × [$\frac{5A}{6}$A])

m_{20} = ([$\frac{5A}{6}$A]× {D}) ∪ ({A} × [O$\frac{A}{3}$])

Each of m_i (1 ≤ i ≤ 20) is a continuum. We designate k_j (1 ≤ j ≤ 10) below as unions of certain m_i containing a common point, hence each k_j is a continuum. Figures 5 and 6 illustrate Z_1.

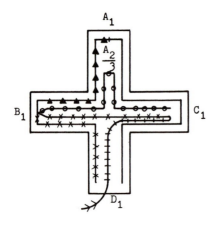
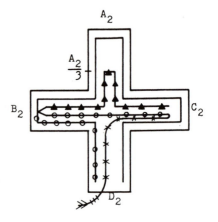
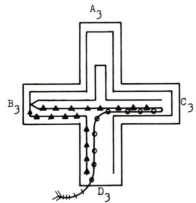
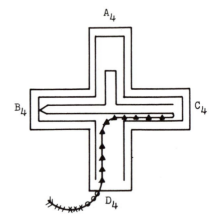

Fig. 4.

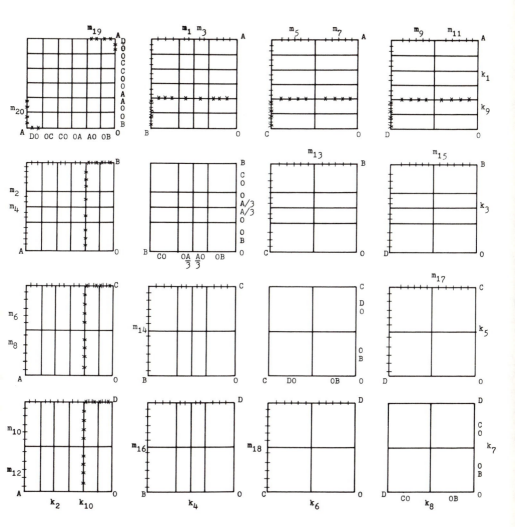

Fig. 5.

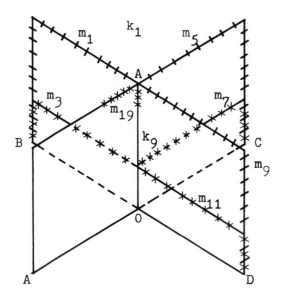

Fig. 6. Part of $X_1 \times X_1$ with part of Z_1 marked.

$k_1 = m_1 \cup m_5 \cup m_9 \cup m_{19}$ $(0,A) \in m_1 \cap m_5 \cap m_9 \cap m_{19}$

$k_2 = m_2 \cup m_6 \cup m_{10} \cup m_{20}$ $(A,0) \in m_2 \cap m_6 \cap m_{10} \cap m_{20}$

$k_3 = m_2 \cup m_4 \cup m_{13} \cup m_{15}$ $(0,B) \in m_2 \cap m_4 \cap m_{13} \cap m_{15}$

$k_4 = m_1 \cup m_3 \cup m_{14} \cup m_{16}$ $(B,0) \in m_1 \cap m_3 \cap m_{14} \cap m_{16}$

$k_5 = m_6 \cup m_8 \cup m_{14} \cup m_{17}$ $(0,C) \in m_6 \cap m_8 \cap m_{14} \cap m_{17}$

$k_6 = m_5 \cup m_7 \cup m_{13} \cup m_{18}$ $(C,0) \in m_5 \cap m_7 \cap m_{13} \cap m_{18}$

$k_7 = m_{10} \cup m_{12} \cup m_{16} \cup m_{18}$ $(0,D) \in m_{10} \cap m_{12} \cap m_{16} \cap m_{18}$

$k_8 = m_9 \cup m_{11} \cup m_{15} \cup m_{17}$ $(D,0) \in m_9 \cap m_{11} \cap m_{15} \cap m_{17}$

$k_9 = m_3 \cup m_7 \cup m_{11}$ $(0,\frac{A}{3}) \in m_3 \cap m_7 \cap m_{11}$

$k_{10} = m_4 \cup m_8 \cup m_{12}$ $(\frac{A}{3},0) \in m_4 \cap m_8 \cap m_{12}$

Observe that: $m_1 \subseteq k_1 \cap k_4$ $m_{16} \subseteq k_4 \cap k_7$ $m_{10} \subseteq k_7 \cap k_2$
$m_6 \subseteq k_2 \cap k_5$ $m_{17} \subseteq k_5 \cap k_8$ $m_{15} \subseteq k_8 \cap k_3$ $m_{13} \subseteq k_2 \cap k_6$.
Hence, $\bigcup_{j=1}^{8} k_j$ is a continuum and further
$(K_9 \cup k_{10}) \subseteq \bigcup_{j=1}^{20} k_j = \bigcup_{i=1}^{20} m_i$. Hence $Z_1 = \bigcup_{i=1}^{20} m_i$ is a con-
tinuum. We further observe that:

For each odd i, $(1 \leq i \leq 19)$, $m_i^{-1} = m_{i+1}$, so $Z_1^{-1} = Z_1$,
$\pi_1(m_2 \cup m_1 \cup m_{13} \cup m_{17}) = \pi_2(m_2 \cup m_1 \cup m_{13} \cup m_{17}) = X_1$,
$(B,C) \in m_{14}$, $(C,B) \in m_{13}$, $(B,A) \in m_1$, $(A,B) \in m_2$,
    $(B,D) \in m_{16}$, $(D,B) \in m_{15}$
If $(p,q) \in Z_1$, then $d(p,q) \geq \frac{1}{3}$.

We adopt the convention that $\langle t,u \rangle$ denotes a continuum M
such that $\pi_1(M) = t$ and $\pi_2(M) = u$ and $\langle t,u \rangle^{-1} = \langle u,t \rangle$.
Note that Z_1 is the union of 20 such continua in the same
order as in the induction hypothesis below. If $\langle t,u \rangle$ is a
subcontinuum of Z_n, and v,w are subarcs of X_1 such that
$f|_v$ maps onto t and $f|_w$ maps onto u, both homeomorphically, then

$L = (f|_v^{-1} \times f|_w^{-1})(\langle t,u \rangle)$ is a continuum such that

$\pi_1(L) = v$ and $\pi_2(1) = w$.

We denote this continuum as $L(\langle t,u \rangle, v, w)$, or more briefly as
L, and call it the *lifting* of $\langle t,u \rangle$ with respect to v and
w. (As defined in [8].)

Induction hypothesis: Z_n is a continuum of $X_1 \times X_1$ such
that
a) $\pi_1(Z_n) = \pi_2(Z_n) = X$.
b) Z_n is the union of twenty continua denoted by:

   $\langle OB,OA \rangle$ $\langle OA,OB \rangle$ $\langle OB,0\frac{A}{3} \rangle$ $\langle 0\frac{A}{3},OB \rangle$

   $\langle OC,OA \rangle$ $\langle OA,OC \rangle$ $\langle OC,0\frac{A}{3} \rangle$ $\langle 0\frac{A}{3},OC \rangle$

   $\langle OD,OA \rangle$ $\langle OA,OD \rangle$ $\langle OD,0\frac{A}{3} \rangle$ $\langle 0\frac{A}{3},OD \rangle$

$$\langle\, OC,OB\,\rangle \quad \langle\, OB,OC\,\rangle \quad \langle\, OD,OB\,\rangle \quad \langle\, OB,OD\,\rangle$$

$$\langle\, OD,OC\,\rangle \quad \langle\, OC,OD\,\rangle \quad \langle\, 0\tfrac{A}{3},\tfrac{5A}{6}A\,\rangle \quad \langle\, \tfrac{5A}{6}A,0\tfrac{A}{3}\,\rangle$$

where $\langle t,u \rangle$ is a continuum $M \ni \pi_1(M) = t$ and $\pi_2(M) = u$.

c) $\langle t,u \rangle^{-1} = \langle u,t \rangle$

d) There are five points in X_1, Z_n as follows:

$x_1 \in \tfrac{5A}{6}A$, with $(x_1,0) \in \langle\, OA,OB\,\rangle \cap \langle\, OA,OC\,\rangle \cap \langle\, OA,OD\,\rangle \cap \langle\, \tfrac{5A}{6}A,0\tfrac{A}{3}\,\rangle$

   and $(0,x_1) \in \langle\, OB,OA\,\rangle \cap \langle\, OC,OA\,\rangle \cap \langle\, OD,OA\,\rangle \cap \langle\, 0\tfrac{A}{3},\tfrac{5A}{6}A\,\rangle$

$x_2 \in \tfrac{2B}{3}B$, with $(x_1,0) \in \langle\, OB,OA\,\rangle \cap \langle\, OB,OC\,\rangle \cap \langle\, OB,OD\,\rangle \cap \langle\, OB,0\tfrac{A}{3}\,\rangle$

   and $(0,x_2) \in \langle\, OA,OB\,\rangle \cap \langle\, OC,OB\,\rangle \cap \langle\, OD,OB\,\rangle \cap \langle\, 0\tfrac{A}{3},OB\,\rangle$

$x_3 \in \tfrac{C}{2}C$, with $(x_2,0) \in \langle\, OC,OA\,\rangle \cap \langle\, OC,OB\,\rangle \cap \langle\, OC,OD\,\rangle \cap \langle\, OC,0\tfrac{A}{3}\,\rangle$

   and $(0,x_3) \in \langle\, OA,OC\,\rangle \cap \langle\, OB,OC\,\rangle \cap \langle\, OD,OC\,\rangle \cap \langle\, 0\tfrac{A}{3},OC\,\rangle$

$x_4 \in \tfrac{D}{2}D$, with $(x_4,0) \in \langle\, OD,OA\,\rangle \cap \langle\, OD,OB\,\rangle \cap \langle\, OD,OC\,\rangle \cap \langle\, OD,0\tfrac{A}{3}\,\rangle$

   and $(0,x_4) \in \langle\, OA,OD\,\rangle \cap \langle\, OB,OD\,\rangle \cap \langle\, OC,OD\,\rangle \cap \langle\, 0\tfrac{A}{3},OD\,\rangle$

$x_5 \in \tfrac{A}{6}\tfrac{A}{3}$, with $(x_5,0) \in \langle\, 0\tfrac{A}{3},OB\,\rangle \cap \langle\, 0\tfrac{A}{3},OC\,\rangle \cap \langle\, 0\tfrac{A}{3},OD\,\rangle$

   and $(0,x_5) \in \langle\, OB,0\tfrac{A}{3}\,\rangle \cap \langle\, OC,0\tfrac{A}{3}\,\rangle \cap \langle\, OD,0\tfrac{A}{3}\,\rangle$

e) There are three points in X_1, Z_n as follows:

$z_1 \in OC$ with $(B,z_1) \in \langle\, OB,OC\,\rangle$ and $(z_1,B) \in \langle\, OC,OB\,\rangle$

$z_2 \in OA$ with $(B,z_2) \in \langle\, OB,OA\,\rangle$ and $(z_2,B) \in \langle\, OA,OB\,\rangle$

$z_3 \in OD$ with $(B,z_3) \in \langle\, OB,OD\,\rangle$ and $(z_3,B) \in \langle\, OD,OB\,\rangle$

Base Case: Observe that Z_1 meets all of the above conditions.

Our induction will be to construct Z_{n+1} by lifting Z_n, so Z_{n+1} satisfies (a)–(e) and $f \times f(Z_{n+1}) = Z_n$. a_i ($1 \leq i \leq 20$) will denote the twenty continua whose union is Z_{n+1}. Furthermore, $\langle t,u \rangle'$ will denote a continuum which "corresponds" to $\langle t,u \rangle$ of Z_n. This proof closely follows Ingram's method in [8], as does the inductive hypothesis above.

$a_1 = \langle OB, OA \rangle'$

$ = L_1^1(\langle OC, OB \rangle, \frac{2B}{3}B, 0\frac{A}{6} \quad) \cup L_2^1(\langle OC, OA \rangle, \frac{2B}{3}B, \frac{A}{6}\,\frac{A}{3} \quad)$

$ \cup L_3^1(\langle 0\frac{A}{3}, \frac{5A}{6}A \rangle, \frac{B}{2}\,\frac{2B}{3}, \frac{11A}{36}\,\frac{A}{3}) \cup L_4^1(\langle 0\frac{A}{3}, \frac{5A}{6}A \rangle, \frac{B}{3}\,\frac{B}{2}, \frac{11A}{36}\,\frac{A}{3})$

$ \cup L_5^1(\langle OB, OA \rangle, 0\frac{B}{3}, \frac{A}{6}\,\frac{A}{3} \quad) \cup L_6^1(\langle OB, OA \rangle, 0\frac{B}{3}, \frac{A}{3}\,\frac{A}{2} \quad)$

$ \cup L_7^1(\langle OB, OC \rangle, 0\frac{B}{3}, \frac{A}{2}\,\frac{2A}{3} \quad) \cup L_8^1(\langle OB, OC \rangle, 0\frac{B}{3}, \frac{2A}{3}\,\frac{5A}{6})$

$ \cup L_9^1(\langle OB, OD \rangle, 0\frac{B}{3}, \frac{5A}{6}A \quad)$

$a_2 = \langle OA, OB \rangle' = a_1^{-1}$

That a_1 is a continuum follows from:

$(x_3, 0) \in \langle OC, OB \rangle \cap \langle OC, OA \rangle$, so $(f^{-1}_{|\frac{2B}{3}B}(x_3), \frac{A}{6}) \in L_1^1 \cap L_2^1$

$(0, x_1) \in \langle OC, OA \rangle \cap \langle 0\frac{A}{3}, \frac{5A}{6}A \rangle$, so $(\frac{2B}{3}, f^{-1}_{|\frac{11A}{36}\frac{A}{3}}(x_1)) \in L_2^1 \cap L_3^1$

$\exists y \in \frac{5A}{6}, \; (\frac{A}{3}, y) \in \langle 0\frac{A}{3}, \frac{5A}{6}A \rangle$, so $(\frac{B}{2}, f^{-1}_{|\frac{11A}{36}\frac{A}{3}}(y)) \in L_3^1 \cap L_4^1$

$(0, x_1) \in \langle 0\frac{A}{3}, \frac{5A}{6}A \rangle \cap \langle OB, OA \rangle$, so $(\frac{B}{3}, f^{-1}_{|\frac{11A}{36}\frac{A}{3}}(x_1)) \in L_4^1 \cap L_5^1$

$\exists y \in OB, \; (h, A) \in \langle OB, OA \rangle$, so $(f^{-1}_{|0\frac{B}{3}}(y), \frac{A}{3}) \in L_5^1 \cap L_6^1$

$(x_2, 0) \in \langle OB, OA \rangle \cap \langle OB, OC \rangle$, so $(f^{-1}_{|0\frac{B}{3}(x_2)}, \frac{A}{2}) \in L_6^1 \cap L_7^1$

$\exists y \in OB, \; (y_1, C) \in \langle OB, OC \rangle$, so $(f^{-1}_{|0\frac{B}{3}(y)}, \frac{2A}{3}) \in L_7^1 \cap L_8^1$

$(x_1, 0) \in \langle OB, OC \rangle \cap \langle OB, OD \rangle$, so $(f^{-1}_{|0\frac{B}{3}(x_2)}, \frac{5A}{6}) \in L_8^1 \cap L_9^1$

So a_2 is also a continuum.

Also: $\pi_1(a_1) = OB$ & $\pi_2(a_1) = OC$ & $f \times f \, (a_1) \subseteq Z_n$

$ \pi_1(a_2) = OC$ & $\pi_2(a_1) = OB$ & $f \, \beta \, f \, (a_2) \subseteq Z_n$

$a_3 = \langle OB, 0\frac{A}{3} \rangle' = \bigcup_{i=1}^{5} L_i^3$ where $L_i^3 = L_i^1$ for $1 \le i \le 5$.

$a_4 = \langle 0\frac{A}{3}, OB \rangle' = a_3^{-1}$

The proof that $a_1(a_2)$ is a continuum contains the proof that $a_3(a_4)$ is a continuum.

Further: $\pi_1(a_3) = OB$ & $\pi_2(a_3) = O\frac{A}{3}$ & $f \times f \ (a_3) \subseteq Z_n$

$\qquad\qquad \pi_1(a_4) = O\frac{A}{3}$ & $\pi_2(a_4) = OB$ & $f \times f \ (a_4) \subseteq Z_n$

$a_5 = \langle OC, OA \rangle' = L_1^5(\langle OD, OB \rangle, \frac{C}{2}C, O\frac{A}{6}) \cup L_2^5(\langle OD, OA \rangle, \frac{C}{2}C, \frac{A}{6}\frac{A}{3})$

$\qquad\qquad \cup L_3^5(\langle OB, OA \rangle, O\frac{C}{2}, \frac{A}{6}\frac{A}{3}) \cup L_4^5(\langle OB, OA \rangle, O\frac{C}{2}, \frac{A}{3}\frac{A}{2})$

$\qquad\qquad \cup L_5^5(\langle OB, OC \rangle, O\frac{C}{2}, \frac{A}{2}\frac{2A}{3}) \cup L_6^5(\langle OB, OC \rangle, O\frac{C}{2}, \frac{2A}{3}\frac{5A}{6})$

$\qquad\qquad \cup L_7^5(\langle OB, OD \rangle, O\frac{C}{2}, \frac{5A}{6}A)$

$a_6 = \langle OA, OC \rangle' = a_5^{-1}$

That a_5 is a continuum follows from:

$(x_4, 0) \in \langle OD, OB \rangle \cap \langle OD, OA \rangle$, so $(f^{-1}_{|\frac{C}{2}C}(x_4), \frac{A}{6}) \in L_1^5 \cap L_2^5$

$(0, x_1) \in \langle OD, OA \rangle \cap \langle OB, OA \rangle$, so $(\frac{C}{2}, f^{-1}_{|\frac{A}{6}\frac{A}{3}}(x_1)) \in L_2^5 \cap L_3^5$

$\exists y \in OB, \ (y, A) \in \langle OB, OA \rangle$, so $(f^{-1}_{|O\frac{C}{2}}(y), \frac{A}{3}) \in L_3^5 \cap L_4^5$

$(x_2, 0) \in \langle OB, OA \rangle \cap \langle OB, OC \rangle$, so $(f^{-1}_{|O\frac{C}{2}}(x_2), \frac{A}{2}) \in L_4^5 \cap L_5^5$

$\exists y \in OB, \ (y, C) \in \langle OB, OC \rangle$, so $(f^{-1}_{|O\frac{C}{2}}(y), \frac{2A}{3}) \in L_5^5 \cap L_6^5$

$(x_2, 0) \in \langle OB, OC \rangle \cap \langle OB, OD \rangle$, so $(f^{-1}_{|O\frac{C}{2}}(x_2), \frac{5A}{6}) \in L_6^5 \cap L_7^5$

So a_6 is also a continuum.

Also: $\pi_1(a_5) = OC$ & $\pi_2(a_5) = OA$ & $f \times f \ (a_5) \subseteq Z_n$

$\qquad\qquad \pi_1(a_6) = OA$ & $\pi_2(a_6) = OC$ & $f \times f \ (a_6) \subseteq Z_n$

$a_7 = \langle OC, O\frac{A}{3} \rangle' = \overset{3}{\underset{i=1}{\cup}} L_i^7$ where $L_i^7 = L_i^5$ for $1 \leq i \leq 3$.

$a_8 = \langle O\frac{A}{3}, OC \rangle' = a_7^{-1}$

That $a_7(a_8)$ is a continuum follows from the proof for $a_5(a_6)$.

Further: $\pi_1(a_7) = OC$ & $\pi_2(a_7) = O\frac{A}{3}$ & $f \times f \ (a_7) \subseteq Z_n$

$\qquad\qquad \pi_1(a_8) = O\frac{A}{3}$ & $\pi_2(a_8) = OC$ & $f \times f \ (a_8) \subseteq Z_n$

$$a_9 = \langle\, OD,OA\,\rangle' = L_1^9(\langle\, OC,OB\,\rangle,\ \tfrac{D}{2}D,0\tfrac{A}{6}\ ) \cup L_2^9(\langle\, OC,OA\,\rangle,\ \tfrac{D}{2}D,\tfrac{A}{6}\,\tfrac{A}{3}\ )$$

$$\cup\ L_3^9(\langle\, OB,OA\,\rangle,\ 0\tfrac{D}{2},\tfrac{A}{6}\,\tfrac{A}{3}\ ) \cup L_4^9(\langle\, OB,OA\,\rangle,\ 0\tfrac{D}{2},\tfrac{A}{3}\,\tfrac{A}{2}\ )$$

$$\cup\ L_5^9(\langle\, OB,OC\,\rangle,\ 0\tfrac{D}{2},\tfrac{A}{2}\,\tfrac{2A}{3}) \cup L_6^9(\langle\, OB,OC\,\rangle,\ 0\tfrac{D}{2},\tfrac{2A}{3}\,\tfrac{5A}{6})$$

$$\cup\ L_7^9(\langle\, OB,OD\,\rangle,\ 0\tfrac{D}{2},\tfrac{5A}{6}A)$$

$$a_{10} = \langle\, OA,OD\,\rangle' = a_9^{-1}$$

That a_9 is a continuum follows from:

$(x_3,0) \in \langle\, OC,OB\,\rangle \cap \langle\, OC,OA\,\rangle$, so $(f^{-1}|_{\frac{D}{2}D}(x_3),\tfrac{A}{6}) \in L_1^9 \cap L_2^9$

$(0,x_1) \in \langle\, OC,OA\,\rangle \cap \langle\, OB,OA\,\rangle$, so $(\tfrac{D}{2},f^{-1}|_{\frac{A}{6}\,\frac{A}{3}}(x_1)) \in L_2^9 \cap L_3^9$

$\exists y \in OB,\ (y,A) \in \langle\, OB,OA\,\rangle$, so $(f^{-1}|_{0\frac{D}{2}}(y),\tfrac{A}{3}) \in L_3^9 \cap L_4^9$

$(x_2,0) \in \langle\, OB,OA\,\rangle \cap \langle\, OB,OC\,\rangle$, so $(f^{-1}|_{0\frac{D}{2}}(x_2),\tfrac{A}{2}) \in L_4^9 \cap L_5^9$

$\exists y \in OB,\ (y,C) \in \langle\, OB,OC\,\rangle$, so $(f^{-1}|_{0\frac{D}{2}}(y),\tfrac{2A}{3}) \in L_5^9 \cap L_6^9$

$(x_2,0) \in \langle\, OB,OC\,\rangle \cap \langle\, OB,OD\,\rangle$, so $(f^{-1}|_{0\frac{D}{2}}(x_2),\tfrac{5A}{6}) \in L_6^9 \cap L_7^9$

So a_{10} is also a continuum.

Also: $\pi_1(a_9) = OD$ & $\pi_2(a_9) = OA$ & $f \times f\ (a_9) \subseteq Z_n$

$\pi_1(a_{10}) = OA$ & $\pi_2(a_{10}) = OD$ & $f \times f\ (a_{10}) \subseteq Z_n$

$a_{11} = \langle\, OD,0\tfrac{A}{3}\,\rangle' = \bigcup_{i=1}^{3} L_i^{11}$ where $L_i^{11} = L_i^9$ for $1 \leq i \leq 3$.

$a_{12} = \langle\, 0\tfrac{A}{3},OD\,\rangle = a_{11}^{-1}$

That $a_{11}(a_{12})$ is a continuum follows from proof for $a_9(a_{10})$.

Also: $\pi_1(a_{11}) = OD$ & $\pi_2(a_{11}) = 0\tfrac{A}{3}$ & $f \times f\ (a_{11}) \subseteq Z_n$

$\pi_1(a_{12}) = 0\tfrac{A}{3}$ & $\pi_2(a_{12}) = OD$ & $f \times f\ (a_{12}) \subseteq Z_n$

$$a_{13} = \langle\, OC,OB\,\rangle' = L_1^{13}(\langle\, OD,OB\,\rangle,\ \tfrac{C}{2}C,0\tfrac{B}{3}\ ) \cup L_2^{13}(\langle\, OD,0\tfrac{A}{3}\,\rangle,\ \tfrac{C}{2}C,\tfrac{B}{3}\,\tfrac{B}{2}\ )$$

$$\cup\ L_3^{13}(\langle\, OB,0\tfrac{A}{3}\,\rangle,\ 0\tfrac{C}{2},\tfrac{B}{3}\,\tfrac{B}{2}) \cup L_4^{13}(\langle\, OB,0\tfrac{A}{3}\,\rangle,\ 0\tfrac{C}{2},\tfrac{B}{2}\,\tfrac{2B}{3})$$

$$\cup\ L_5^{13}(\langle\, OB,OC\,\rangle,\ 0\tfrac{C}{2},\tfrac{2B}{3}B)$$

$$a_{14} = \langle OB, OC \rangle = a_{13}^{-1}$$

That a_{13} is a continuum follows from:

$(x_4, 0) \in \langle OD, OB \rangle \cap \langle OD, 0\frac{A}{3} \rangle$, so $(f^{-1}_{|\frac{C}{2}C}(x_4), \frac{B}{3}) \in L_1^{13} \cap L_2^{13}$

$(0, x_5) \in \langle OD, 0\frac{A}{3} \rangle \cap \langle OB, 0\frac{A}{3} \rangle$, so $(\frac{C}{2}, f^{-1}_{|\frac{B}{3}\frac{B}{2}}(x_5)) \in L_2^{13} \cap L_3^{13}$

$\exists y \in OB, \ (y, \frac{A}{3}) \in \langle OB, 0\frac{A}{3} \rangle$, so $(f^{-1}_{|0\frac{C}{2}}(y), \frac{B}{2}) \in L_3^{13} \cap L_4^{13}$

$(x_2, 0) \in \langle OB, 0\frac{A}{3} \rangle \cap \langle OB, OC \rangle$, so $(f^{-1}_{|0\frac{C}{2}}(x_2), \frac{2B}{3}) \in L_4^{13} \cap L_5^{13}$

So a_{14} is also a continuum.

Also: $\pi_1(a_{13}) = OC$ & $\pi_2(a_{13}) = OB$ & $f \times f \ (a_{13}) \subseteq Z_n$

$\pi_1(a_{14}) = OB$ & $\pi_2(a_{14}) = OC$ & $f \times f \ (a_{14}) \subseteq Z_n$

$a_{15} = \langle OD, OB \rangle' = L_1^{15}(\langle OC, OB \rangle, \frac{D}{2}D, 0\frac{B}{3}) \cup L_2^{15}(\langle OC, 0\frac{A}{3} \rangle, \frac{D}{2}D, \frac{B}{3}\frac{B}{2})$

$\cup L_3^{15}(\langle OB, 0\frac{A}{3} \rangle, 0\frac{D}{2}, \frac{B}{3}\frac{B}{2}) \cup L_4^{15}(\langle OB, 0\frac{A}{3} \rangle, 0\frac{D}{2}, \frac{B}{2}\frac{2B}{3})$

$\cup L_5^{15}(\langle OB, OC \rangle, 0\frac{D}{2}, \frac{2B}{3}B)$

$a_{16} = \langle OB, OD \rangle' = a_{15}^{-1}$

That a_{15} is a continuum follows from:

$(x_3, 0) \in \langle OC, OB \rangle \cap \langle OC, 0\frac{A}{3} \rangle$, so $(f^{-1}_{|\frac{D}{2}D}(x_3), \frac{B}{3}) \in L_1^{15} \cap L_2^{15}$

$(0, x_5) \in \langle OC, 0\frac{A}{3} \rangle \cap \langle OB, 0\frac{A}{3} \rangle$, so $(\frac{D}{2}, f^{-1}_{|\frac{B}{3}\frac{B}{2}}(x_5)) \in L_2^{15} \cap L_3^{15}$

$\exists y \in OB, \ (y, \frac{A}{3}) \in \langle OB, 0\frac{A}{3} \rangle$, so $(f^{-1}_{|0\frac{D}{2}}(y), \frac{B}{2}) \in L_3^{15} \cap L_4^{15}$

$(x_2, 0) \in \langle OB, 0\frac{A}{3} \rangle \cap \langle OB, OC \rangle$, so $(f^{-1}_{|0\frac{D}{2}}(x_2), \frac{2B}{3}) \in L_4^{15} \cap L_5^{15}$

So a_{16} is also a continuum.

Also: $\pi_1(a_{15}) = OD$ & $\pi_2(a_{15}) = OB$ & $f \times f \ (a_{15}) \subseteq Z_n$

$\pi_1(a_{16}) = OB$ & $\pi_2(a_{16}) = OD$ & $f \times f \ (a_{16}) \subseteq Z_n$

$a_{17} = \langle OD, OC \rangle' = L_1^{17}(\langle OC, OB \rangle, \frac{D}{2}D, 0\frac{C}{2}) \cup L_2^{17}(\langle OC, OD \rangle, \frac{D}{2}D, \frac{C}{2}C)$

$\cup L_3^{17}(\langle OB, OD \rangle, 0\frac{D}{2}, \frac{C}{2}C)$

$a_{18} = \langle OC, OD \rangle' = a_{17}^{-1}$

That a_{17} is a continuum follows from:

$(x_3, 0) \in \langle OC, OB \rangle \cap \langle OC, OD \rangle$, so $(f^{-1}{\mid}_{\frac{D}{2}D}(x_3), \frac{C}{2}) \in L_1^{17} \cap L_2^{17}$

$(0, x_4) \in \langle OC, OD \rangle \cap \langle OB, OD \rangle$, so $(\frac{D}{2}, f^{-1}{\mid}_{\frac{C}{2}C}(x_4)) \in L_2^{17} \cap L_3^{17}$

So a_{18} is also a continuum.

Also: $\pi_1(a_{17}) = OD$ & $\pi_2(a_{17}) = OC$ & $f \times f (a_{17}) \subseteq Z_n$

$\pi_1(a_{18}) = OC$ & $\pi_2(a_{18}) = OD$ & $f \times f (a_{18}) \subseteq Z_n$

$a_{19} = \langle O\frac{A}{3}, \frac{5A}{6}A \rangle' = L_1^{19}(\langle OA, OD \rangle, \frac{A}{6} \frac{A}{3}, \frac{5A}{6}A) \cup L_2^{19}(\langle OB, OD \rangle, O\frac{A}{6}, \frac{5A}{6}A)$

$a_{20} = \langle \frac{5A}{5}A, O\frac{A}{3} \rangle' = a_{19}^{-1}$

That a_{19} is a continuum follows from:

$(0, x_4) \in \langle OA, OD \rangle \cap \langle OB, OD \rangle$, so $(\frac{A}{6}, f^{-1}{\mid}_{\frac{5A}{6}A}(x_4)) \in L^{19} \cap L^{19}$

So a_{20} is also a continuum.

Also: $\pi_1(a_{19}) = O\frac{A}{3}$ & $\pi_2(a_{19}) = \frac{5A}{6}A$ & $f \times f (a_{19}) \subseteq Z_n$

$\pi_1(a_{20}) = \frac{5A}{6}A$ & $\pi_2(a_{20}) = O\frac{A}{3}$ & $f \times f (a_{20}) \subseteq Z_n$

We now show $Z_{n+1} = \bigcup_{i=1}^{20} a_i$ is a continuum satisfying (a)-(e) of the induction hypothesis and such that $f \times f (Z_{n+1}) = Z_n$. Designate the following five points:

$x_1' = f^{-1}{\mid}_{\frac{5A}{6}A}(z_3)$ & $x_2' = f^{-1}{\mid}_{\frac{2B}{3}B}(z_1)$ & $x_3' = f^{-1}{\mid}_{\frac{C}{2}C}(z_3)$

$x_4' = f^{-1}{\mid}_{\frac{D}{2}D}(z_1)$ & $x_5' = f^{-1}{\mid}_{\frac{A}{6}A}(z_2)$

These points will be shown below to satisfy (d) and will be used to show b_i $(1 \leq i \leq 10)$, defined below, are each continua. (Read "$L_9^2 \to (x_1', 0) \in a_2$" as "Lift L_9^2 implies $(x_1', 0) \in a_2 = \langle OA, OB \rangle'$"):

$b_1 = a_2 \cup a_6 \cup a_{10} \cup a_{20}$, where $L_9^2 \to (x_1', 0) \in a_2$,

$L_7^6 \to (x_1', 0) \in a_6$, $L_7^{10} \to (x_1', 0) \in a_{10}$, & $L_2^{20} \to (x_1', 0) \in a_{20}$.

$b_2 = a_1 \cup a_5 \cup a_9 \cup a_{19}$, where $L_9^1 \to (0, x_1') \in a_1$,
$L_7^5 \to (0, x_1') \in a_5$, $L_7^9 \to (0, x_1') \in a_9$, & $L_2^{19} \to (0, x_1') \in a_{19}$.

$b_3 = a_1 \cup a_{14} \cup a_{16} \cup a_3$, where $L_1^1 \to (x_2', 0) \in a_1$,
$L_5^{14} \to (x_2', 0) \in a_{14}$, $L_5^{16} \to (x_2', 0) \in a_{16}$, & $L_1^3 \to (x_2', 0) \in a_3$.

$b_4 = a_2 \cup a_{13} \cup a_{15} \cup a_4$, where $L_1^2 \to (0, x_2') \in a_2$,
$L_5^{13} \to (0, x_2') \in a_{13}$, $L_5^{15} \to (0, x_2') \in a_{15}$, & $L_1^4 \to (0, x_2') \in a_4$.

$b_5 = a_5 \cup a_{13} \cup a_{18} \cup a_7$, where $L_1^5 \to (x_3', 0) \in a_5$,
$L_1^{13} \to (x_3', 0) \in a_{13}$, $L_3^{18} \to (x_3', 0) \in a_{18}$, & $L_1^7 \to (x_3', 0) \in a_7$.

$b_6 = a_6 \cup a_{14} \cup a_{17} \cup a_8$, where $L_1^6 \to (0, x_3') \in a_6$,
$L_1^{14} \to (0, x_3') \in a_{14}$, $L_3^{17} \to (0, x_3') \in a_{17}$, & $L_1^8 \to (0, x_3') \in a_8$.

$b_7 = a_9 \cup a_{15} \cup a_{17} \cup a_{11}$, where $L_1^9 \to (x_4', 0) \in a_9$,
$L_1^{15} \to (x_4', 0) \in a_{15}$, $L_1^{17} \to (x_1', 0) \in a_{17}$, & $L_1^{11} \to (x_4', 0) \in a_{11}$.

$b_8 = a_{10} \cup a_{16} \cup a_{18} \cup a_{12}$, where $L_1^{10} \to (0, x_4') \in a_{10}$,
$L_1^{16} \to (0, x_4') \in a_{16}$, $L_1^{18} \to (0, x_4') \in a_{18}$, & $L_1^{12} \to (0, x_4') \in a_{12}$.

$b_9 = a_4 \cup a_8 \cup a_{12}$, where $L_5^4 \to (x_5', 0) \in a_4$, $L_3^8 \to (x_5', 0) \in a_8$,
& $L_3^2 \to (x_5', 0) \in a_{12}$.

$b_{10} = a_3 \cup a_7 \cup a_{11}$, where $L_5^3 \to (0, x_5') \in a_3$, $L_3^7 \to (0, x_5') \in a_7$,
& $L_3^{11} \to (0, x_5') \in a_{11}$.

So $x_1', x_2', x_3', x_4', x_5'$ satisfy (d). Observe that:
$a_2 \subseteq b_1 \cap b_4$, $a_{13} \subseteq b_4 \cap b_5$, $a_{18} \subseteq b_5 \cap b_8$, $a_{12} \subseteq b_8 \cap b_9$,
$a_8 \subseteq b_9 \cap b_6$, $a_{14} \subseteq b_6 \cap b_3$, $a_1 \subseteq b_3 \cap b_2$, $a_9 \subseteq b_2 \cap b_7$, and
$a_{11} \subseteq b_7 \cap b_{10}$. Hence $Z_{n+1} = \bigcup_{i=1}^{20} a_i$ is a continuum, and is
the union of twenty continua satisfying (b).

Note that $\pi_1(a_1 \cup a_{13} \cup a_{17} \cup a_{10}) = \pi_2(a_1 \cup a_{13} \cup a_{17} \cup a_{10}) = X_1$, so (a) is also satisfied by Z_{n+1}. Furthermore, (e) is satisfied since:

$\exists z_1' \in OC$, $(B,z_1') \in a_{14} = \langle OB,OC \rangle'$ & $(z_1',B) \in a_{13} = \langle OC,OB \rangle'$

$\exists z_2' \in OA$, $(B,z_2') \in a_1 = \langle OB,OA \rangle'$ & $(z_2',B) \in a_2 = \langle OA,OB \rangle'$

$\exists z_3' \in OD$, $(B,z_3') \in a_{16} = \langle OB,OD \rangle'$ & $(z_3',B) \in a_{15} = \langle OD,OB \rangle'$

As $a_i^{-1} = a_{i+1}$ for odd i, $1 \le i \le 19$, (c) is satisfied by Z_{n+1}.

Finally, we must show $f \times f (Z_{n+1}) = Z_n$. Since $f \times f (a_i) \subseteq Z_n$ for $1 \le i \le 20$, we need only show $Z_n \subseteq f \times f (Z_{n+1})$. We observe that $f \times f (a_3 \cup a_{13} \cup a_{15} \cup a_{18} \cup a_{20})$ includes:

$\langle OB,OA \rangle$, $\langle OB,0\frac{A}{3} \rangle$, $\langle OC,OA \rangle$, $\langle OC,0\frac{A}{3} \rangle$, $\langle OD,OA \rangle$

$\langle OD,0\frac{A}{3} \rangle$, $\langle OC,OB \rangle$, $\langle OD,OB \rangle$, $\langle OD,OC \rangle$, $\langle 0\frac{A}{3},\frac{5A}{6}A \rangle$.

Hence $f \times f (a_4 \cup a_{14} \cup a_{16} \cup a_{17} \cup a_{19})$ includes:

$\langle OA,OB \rangle$, $\langle 0\frac{A}{3},OB \rangle$, $\langle OA,OC \rangle$, $\langle 0\frac{A}{3},OC \rangle$, $\langle OA,OD \rangle$

$\langle 0\frac{A}{3},OD \rangle$, $\langle OB,OC \rangle$, $\langle OB,OD \rangle$, $\langle OC,OD \rangle$, $\langle \frac{5A}{6}A,0\frac{A}{3} \rangle$.

Thus $Z_n \subseteq f \times f (Z_{n+1})$. Therefore, $f \times f (Z_{n+1}) = Z_n$. Our inductive step is thereby completed. Since $(p,q) \in Z_1$ implies $d(p,q) > \frac{1}{3}$, it follows that $\sigma f^n \ge \frac{1}{3}$, for all n and the theorem is proved.   QED

Figure 5 represents $Z_1 \subseteq X_1 \times X_1$ with hatched and x-ed lines.

Figure 7 represents $Z_2 \subseteq X_1 \times X_1$ with hatched lines going to hatched lines and x-ed lines going to x-ed lines in Figure 5 under $f \times f$.

Two parts of Z_3 are illustrated at the bottom of Figure 7.

By Theorem 2.6, X satisfies the conditions of Theorem 2.5, hence we may immediately conclude:

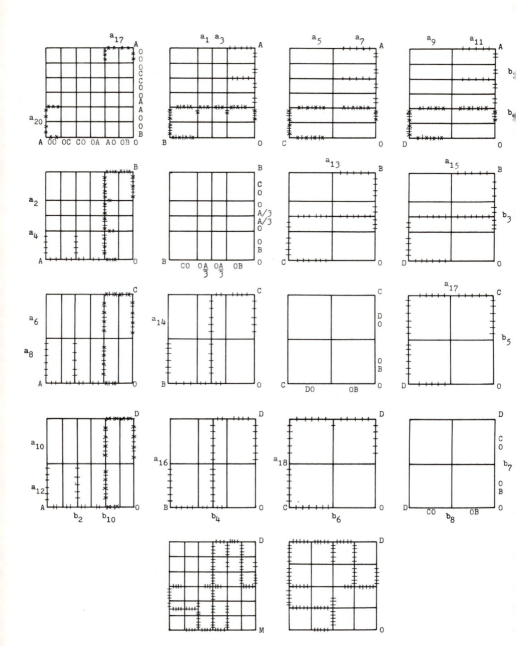

Fig. 7.

2.7. <u>Theorem</u>.

X is nonchainable.

That X is nonseparating is evident. That X is indecomposable is a corollary to the proof of Theorem 2.3 by applying
Ingram and Cook's criterion of indecomposability [9]. Another
proof that X is indecomposable follows from Theorem 3.1 and
a theorem of Sieklucki [14] quoted in [5, Lemma 2.2]. It is
evident from the construction of X that each proper subcontinuum, being chainable, is also an arc, since all subchains
are relatively straight in the covers they refine.

We observe that X has two and only two endpoints, e and
f, such that e is the intersection of the tower:

$$\cdots \subseteq D_5 \subseteq C_4 \subseteq D_3 \subseteq C_2 \subseteq D_1$$

and f is the intersection of the tower:

$$\cdots \subseteq C_5 \subseteq D_4 \subseteq C_3 \subseteq D_2 \subseteq C_1$$

3. LAKE-OF-WADA CHANNELS

By Lake-of-Wada channel we mean a *simple canal*, defined in [5],
which definition is due to Sieklucki [14]. (For the original
Lake-of-Wada construction see [6, pp. 143-144].)

3.1. <u>Definition</u>.

(Sieklucki [14].) Let X be a nonseparating continuum and let
D be a set homeomorphic to [0,1) in $S^2 - X$, where
$\alpha : D \to [0,1)$ is a given homeomorphism. Then D will be
called a *simple canal in* X iff the following three conditions are satisfied:

   (1) $\overline{D} - D \subseteq Bd\ X$

   (2) For each $p \in D$, there is a "bridge" to X; that is
       a crosscut to X, which (crosscut) is transverse to
       D, and intersects D at exactly one point.

(3) If $p_i \to \infty$ (i.e., $\alpha(p_i) \to 1$), then there is a sequence
of bridges $\{Q_{p_i}\}$ such that $Q_{p_i} \cap D = \{p_i\}$ and
diam $Q_{p_i} \to 0$.

If, in addition, condition (4) holds, where (4) is:

(4) $\overline{D} - D = $ Bd X,

we call D a *simple dense canal (s-d-c)*.

We can show that the embedding of X given in 2.2 has a
s-d-c or Lake-of-Wada channel in either of two ways: by directly constructing the requisite ray in $E^2 - X$, or by defining a chain of crosscuts $\{Q_i\}_{i=1}^{\infty}$ of $E^2 - X$ that defines
a prime end E of the third kind with $I(E) = P(E) = X$. (See
Brechner [4] for basic definition of prime ends.) For the
equivalence of these methods see [5, Th. 2.9]. An embedding
for which there exists a s-d-c D in $E^2 - X$ such that
$\overline{D} - D = $ Bd X, is termed *principal* in [5, Definition 2.6].
The embedding of X here given is consequently principal.

The first four stages of the construction of a s-d-c D are
illustrated in Figure 4. Figure 2 illustrates two stages of
D (the "railroad tracks"). Alternately, let Q_i be a cross-
cut of $E^2 - X$ such that for all $i \geq 2$, the endpoints of
Q_i lie in C_i and A_i, while $Q_i \subseteq D_{i-1}$. Then either the
odd or the even subsequence of $\{Q_i\}_{i=2}^{\infty}$ defines a prime end
E such that $I(E) = P(E) = K$. For the even subsequence,
$Q_i \to e$, and for the odd subsequence, $Q_i \to f$, where e and
f are the endpoints of X. Consequently,

3.2 Theorem.

X can be embedded with a simple dense canal D in $E^2 - X$
such that $\overline{D} - D = X$. That is, the embedding of 2.2 is
principal.

3.3 Corollary.

X is indecomposable.

Proof. Follows from 3.1 and lemma 2.2 of [5].

If every embedding of X into the plane were such that E^2 - X contained a s-d-c, then X would be a *principal* continuum in the sense of definition 2.6 of [5]. As a principal continuum X would be a candidate for a nonseparating plane continuum admitting a fixed point free map. Thus two questions suggest themselves:

3.4 Question.

Is X a principal continuum?

3.5 Question.

Does X have the fixed point property for continuous maps?

In [11] Lelek asks if there is an example of a nonchainable continuum with span zero. This author, in efforts to modify Ingram's example to produce a principally embedded, atriodic, nonchainable continuum considered several examples for which said author was not able to prove the span nonzero. On the other hand, the examples are principally embedded, and *appear* to be nonchainable, though atriodic. We show a schematic diagram for the bonding function for one such example as an inverse limit of T's in Figure 8. (For surjective span see Lelek [12].)

3.6 Question.

Does the continuum of Figure 8 have a span greater than zero? Surjective span greater than zero?

3.7 Question.

Is the continuum of Figure 8 chainable?

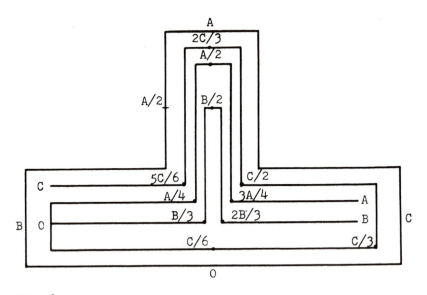

Fig. 8.

REFERENCES

[1] H. Bell, "On fixed point properties of plane continua,"
     Transactions of AMS 128 (1967), 539-548.

[2] D. P. Bellamy, "A tree-like continuum without the fixed
     point property," preprint.

[3] B. Brechner, "On stable homeomorphisms and imbeddings of
     the pseudo arc," Illinois Journal of Mathematics 22
     (1978), 630-661.

[4] ------, "Prime ends, indecomposable continua, and the
     fixed point property," Topology Proceedings, Ohio Uni-
     versity, Conference, March 1979.

[5] B. Brechner and J. C. Mayer, The prime end structure of
     indecomposable continua and the fixed point property,
     Conference in honor of F. B. Jones, University of
     California, Riverside, May 1980.

[6] J. G. Hocking and G. S. Young, Topology, Addison-Wesley,
     Reading, Mass., 1961.

[7] W. T. Ingram, "Decomposable circle-like continua,"
     Fundamenta Mathematica 63 (1968), 193-198.

[8] ------, "An atriodic tree-like continuum with positive
     span," Fundamenta Mathematica 77 (1972), 99-107.

[9] W. T. Ingram and H. Cook, "A characterization of decom-
     posable continua," preprint.

[10] A Lelek, "Disjoint mappings and the span of spaces,"
     Fundamenta Mathematica 63 (1968), 199-214.

[11] ------, (1975). (See University of Houston Problem Book,
     problems 81 and 59.)

[12] ------, "On the surjective span and semispan of connected
     metric spaces," Colloquium Mathematicum 37 (1977), 35-45.

[13] L. G. Oversteegen and James T. Rogers, "Fixed-point-free
     maps on tree-like continua," preprint.

[14] K. Sieklucki, "On a class of plane acyclic continua with
     the fixed point property," Fundamenta Mathematica 63
     (1968), 257-278.

When Homogeneous Continua Are Hausdorff Circles (or Yes, We Hausdorff Bananas)

Forest W. Simmons

In his February 1949 Bulletin article on *Planar Homogeneous Continua*, F. Burton Jones showed that among all homogeneous metric continua, only the simple closed curve is separated by some pair of points.

The purpose of the present paper is to show that the following hausdorff analog of that theorem is also true:

MAIN THEOREM: If a homogeneous hausdorff continuum is separated by some pair of points, then it is separated by each pair of its points, i.e. it is a *hausdorff circle*.

Strategy of Proof. To prove this theorem, first we show that (in the hypothesized space) some nested collection of two-point-boundary subcontinua zeroes in on a point. Then we show that each point separates small, connected, two-point-boundary, open sets. Finally, we show that each pair of points bounds a simple chain of two-point-boundary subcontinua whose union is another such subcontinuum.

Before embarking on this program of proof, let us take the precaution of saying what meaning we assign to certain words and abbreviations:

A *continuum* is a compact connected hausdorff space.

A *separation* is a partition of a (subset of a) space into non-empty (relatively) open sets.

A set Y *separates* the set X whenever X - Y is not connected.

A set N is a *neighborhood* of a point p if p is in
some open subset of N.

The sets *Bd Y*, *Int Y*, and *Cl Y* are the topological
boundary, interior, and closure of Y.

Finally (and best of all) a *banana* is a subcontinuum with
two point boundary.

Now that we know what a banana is, let us establish some
lemmas (lemons?). The first five lemmas are basic facts which
we will use several times in the rest of the paper, especially
in Lemma 6, the crucial lemma in carrying out the first part
of our program of proof as outlined above.

LEMMA 1: If X is a homogeneous continuum and A | B is
a separation of X - {x,y}, then A ∪ {x,y} and B ∪ {x,y}
are bananas with common boundary {x,y}.

Proof. The set X - {x} is connected since no singleton
separates a homogeneous continuum. The set A ∪ {y} is connected since the connected set {y} separates the connected
set X - {x} into mutually separated sets A and B. Similarly, A ∪ {x} is connected. So A ∪ {x,y} is connected.
Furthermore, A ∪ {x,y} has no limit points in its complement
B, so A ∪ {x,y} is a subcontinuum. Similarly, B ∪ {x,y}
is a subcontinuum. The two point set {x,y} is the common
boundary of these subcontinua.

COROLLARY 2: If X is a homogeneous continuum and H is
a banana of X, then Cl(X - H) is also a banana.

Proof. Let A = Int H, B = X - H, and {x,y} = Bd H,
and apply Lemma 1.

LEMMA 3: Let H be a banana of a homogeneous continuum X, and let p be a point in X. If $A_1 \mid A_2$ is a separation of $H - \{p\}$, then each A_i intersects the boundary of H in one point, and each A_i is connected. In particular, if $\{p\}$ separates H, then $p \in$ Int H.

Proof. Any A_i which missed Bd H would induce a separation of $X - \{p\}$, a connected set. But Bd H has only two points, so each A_i has exactly one of them. Any disconnected A_i would split into two non-empty mutually separated sets, one of which would miss Bd H, thereby giving rise to a separation of $X - \{p\}$.

LEMMA 4: Let H be a banana of a homogeneous continuum X, let $K = Cl(X - H)$, let $x \in$ Int H, and let $y \in$ Int K. If $\{x,y\}$ separates X, then $\{x\}$ separates H.

Proof. First note that K is a banana by Corollary 2. Then let $A = (K - \{y\}) \cup (H - \{x\})$. If $H - \{x\}$ is connected, then A is connected because (by Lemma 3) each component of $K - \{y\}$ has a point in Bd K, which is contained in $H - \{x\}$ since $x \in$ Int H. But $A = X - \{x,y\}$ since $y \notin H$ and $x \notin K$. In summary, if $\{x\}$ does not separate H, then $\{x,y\}$ does not separate X.

LEMMA 5: Let X be a homogeneous continuum containing a banana H having boundary $\{a,b\}$, and let x be a point of X. If $H - \{x\}$ is not connected, then H is the union of two bananas H_1 and H_2 with boundaries $\{a,x\}$ and $\{x,b\}$ respectively.

Proof. Apply Lemma 3, then Lemma 1.

Armed with some simple facts about bananas in homogeneous
continua, we now attack Lemma 6 and its corollaries which are
crucial for understanding the topology of these homogeneous
banana spaces.

LEMMA 6:   Let H be a non-empty nested collection of
bananas in a homogeneous continuum X. If $\cap H$ is not a singleton, then there is a banana which is properly contained in
every banana in the collection H.

Proof.   Let $K = \{Cl(X - H) \mid H \in H\}$, the collection of
bananas complementary to those of H as provided by Corollary
2.

Claim.   There is a pair $(p,q) \in X^2$ such that $\{p,q\}$
separates X, and for each $H \in H$, $p \in$ Int H and $q \in$ H.
Before we prove the claim, let us see how it implies the
lemma:
Suppose (p,q) satisfies the claim. Let $A \mid B$ be a separation of $X - \{p,q\}$. For each $K \in K$, the point q is
not interior to K, so (by Lemma 3) the set $K - \{q\}$ is
connected. Then $\cup K - \{q\}$ is a connected subset of
$X - \{p,q\}$. Without loss, $\cup K - \{q\} \subset B$. Then $A \subset \cap H$.
But $\{p,q\} \subset \cap H$, so $\cap H \supset A \cup \{p,q\}$, a banana by Lemma
1. Furthermore, for each $H \in H$, the bananas H and
$A \cup \{p,q\}$ are unequal since the point p is in the interior of one of them and in the boundary of the other. The
claim is indeed sufficient to prove the lemma.

Proof of Claim.   First pick distinct x and x' in $\cap H$
such that neither x nor x' is a boundary point of any
H in H. This choice is possible since $\cap H$ is a

continuum with at most two points in the boundary of some
H in H. Next, pick y and y' in X so that both
{x,y} and {x',y'} separate X. These choices are possible since X is a homogeneous space containing a banana.
The pair (p,q) is chosen according to three cases:

(1)   If y $\in \cap H$, then take (p,q) = (x,y).

(2)   If y $\notin \cap H$ but y' $\in \cap H$, then take (p,q) = (x',y').

(3)   If neither y nor y' is in $\cap H$, then take
      (p,q) = (x,x').

Finally, we show that in case (3) the pair {x,x'} separates X. To this end, pick H $\in H$ such that
{y,y'} $\cap$ H = ϕ, and let K be the corresponding banana
in K. Then {y,y'} $\subset$ Int K and {x,x'} $\subset$ Int H as in
Fig. 1. Apply Lemma 4 to show that {x} separates H.
Apply Lemma 5 to show that x' is in a banana H' $\subset$ H
such that Bd H' = {x,z} for some z $\in$ Bd H. Apply Lemma
4 again to show that {x'} separates H'. Apply Lemma 5
again to show that {x,x'} is the boundary of some banana
in X.

COROLLARY 7: If X is a homogeneous continuum containing
a banana, then some (hence every) singleton of X is the intersection of a nested collection of bananas.

Proof. Partially order the collection of bananas of X by
containment. Let H be a maximal chain. Then $\cap H$ is a singleton by Lemma 6.

COROLLARY 8: If X is a homogeneous continuum containing
a banana, then every open subset of X contains a banana of
X.

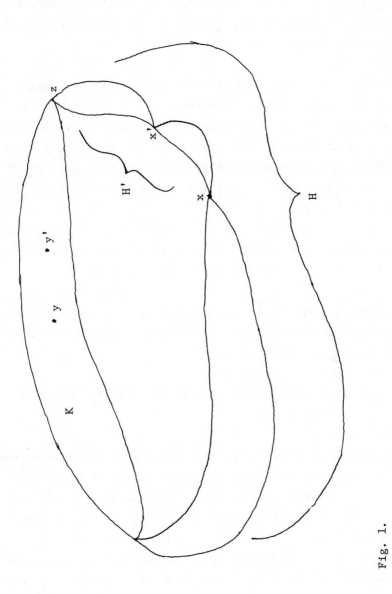

Fig. 1.

Proof. Apply Corollary 7 to any point of the hypothesized open set (and remember that bananas are compact).

COROLLARY 9: If X is a homogeneous continuum containing a banana, then some (hence every) point of X has a basis of banana neighborhoods.

Proof. Partially order the collection of bananas again, but this time by *interior* containment. Let H be a maximal chain. Then $\cap H$ is a singleton. Otherwise, $\cap H$ would contain a banana (by Lemma 6) with another banana in its interior (by Corollary 8), thereby contradicting the maximality of H. Let $\{p\} = \cap H$. Then H is a basis of banana neighborhoods for the point p.

We are now in a position to give a description of some small neighborhoods in "homogeneous banana spaces."

LEMMA 10: Let X be a homogeneous continuum containing a banana. Then each point $p \in X$ has a neighborhood basis B such that for each $B \in B$
  (i)   Int B is connected, and
  (ii) B is the union of bananas B_1 and B_2 such that
        $B_1 \cap B_2 = \{p\}$.

Proof. Some point of X is in the boundary of some banana, so by homogeneity the point p is in Bd H for some banana H. Let q be the other boundary point (as in Figure 2) and let N be any neighborhood of p. Use Corollary 9 to get a banana neighborhood B of p such that $B \subset N - q$. Let $\{b_1, b_2\} = Bd\ B$. Use Lemma 4 to show that $\{p\}$ separates B. Use Lemma 5 to show that B is the union of bananas B_1 and

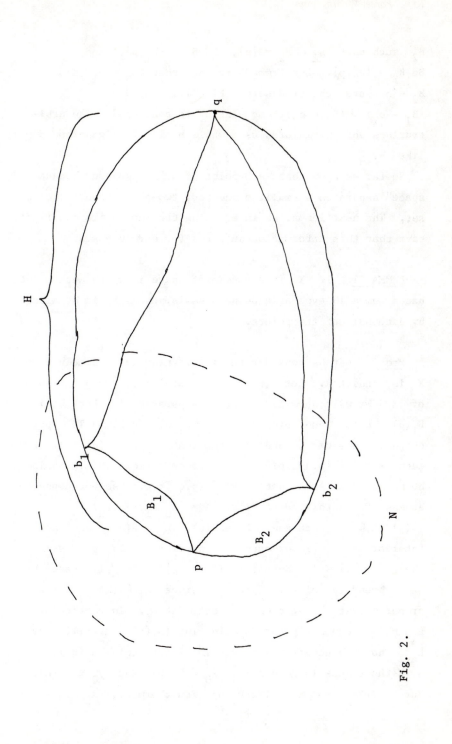

Fig. 2.

B_2 such that $B_1 \cap B_2 = \{p\}$, $Bd\ B_1 = \{b_1, p\}$, and $Bd\ B_2 = \{b_2, p\}$. Use Lemma 3 to show that $B_1 - b_1$ and $B_2 - b_2$ are connected sets. Finally, $Int\ B = (B_1 - b_1) \cup (B_2 - b_2)$, so $Int\ B$ is connected. The arbitrariness of N insures us a whole basis of neighborhoods like B.

So far we know that each point of a "homogeneous banana space" separates a small, connected, two-point-boundary open set. The next lemma, an interesting theorem in its own right, says that this information suffices for our purposes.

LEMMA 11: If X is a hausdorff space in which each point has a Lemma 10 type neighborhood basis, then X is separated by each pair of its points.

Proof. Let us consider the interesting case, namely when X is connected. Let $\{p, q\}$ be an arbitrary two point subset of X. We will show that $\{p, q\}$ separates X. Let $B(p) = B_1(p) \cup B_2(p)$ and $B(q) = B_1(q) \cup B_2(q)$ be Lemma 10 type neighborhoods of p and q such that $B(p) \cap B(q) = \phi$. For each $x \in X - \{p, q\}$ pick $B(x)$ to be a Lemma 10 type neighborhood of X such that $B(x) \cap \{p, q\} = \phi$. Use the connectedness of X to find (finite) $\lambda = \{L_0, L_1, \cdots, L_n\} \subset \{B(x) \mid x \in X\}$ such that $L_0 = B(p)$, $L_n = B(q)$, and the interior of L_i intersects the interior of L_j precisely when $|i - j| \leq 1$. Note that if $0 < |i - j| \leq 1$, then L_i has a boundary point in $Int\ L_j$, since $L_i \cap Int\ L_j$ is a proper subset of the connected set $Int\ L_j$. In particular, L_1 has a boundary point x_0 in $Int\ (B_1(p) \cup B_2(p))$, and L_{n-1} has a boundary point y_0 in $Int\ (B_1(q) \cup B_2(q))$.

Without loss in generality $x_0 \in B_1(p)$ and $y_0 \in B_1(q)$. But $x_0 \neq p$, so $x_0 \in Int\ B_1(p)$. Furthermore, x_0 is a

limit point of $\text{Int } L_1$, so $\text{Int } B_1(p) \cap \text{Int } L_1$ is non-empty.
Then $B_1(p) \cap \text{Int } L_1$ is a proper subset of the connected set
$\text{Int } L_1$, so $B_1(p)$ has a boundary point in $\text{Int } L_1$. Similarly,
$(\text{Bd } B_1(q)) \cap \text{Int } L_{n-1}$ is non-empty.

Let $\beta = \{B_1(p), L_1, \cdots L_{n-1}, B_1(q)\}$. *So far we have shown
that both boundary points of each member of β are interior
to other members of β with the exception of p and q.*
So $\cup\beta - \{p,q\}$ is both open and closed in $X - \{p,q\}$. To
finish the proof we only need to show that $\cup\beta - \{p,q\}$ is a
proper subset of $X - \{p,q\}$. This proper containment will be
verified by showing that $B_2(p) - p$ is not contained in
$\cup\beta$: First, $B_1(p) \cap (B_2(p) - p) = \phi = B_1(q) \cap B_2(p)$. Next,
if $K = \cup\{L_i \mid 0 < i < n\}$, then K is a compact set missing
p, and therefore p is not a limit point of K. But p is
a limit point of $B_2(p) - p$, so this last set is not contained in K.

In summary, $B_2(p) - p$ is not contained in
$K \cup B_1(p) \cup B_1(q) = \cup\beta$, so $(\cup\beta - \{p,q\}) \mid (X - \cup\beta)$ is a
genuine separation of $X - \{p,q\}$. We are now in a position
to prove the main theorem.

Proof of Main Theorem. Let X be a homogeneous continuum
which is separated by some two point subset. Apply Lemma 1 to
show that X has a banana. Then show that X is separated
by each two point subset by applying Lemmas 10 and 11.

Remarks. My first attempt at proving the above theorem
made use of a certain "Weak Effros Property" which provides
space homeomorphisms that jiggle one point a little without
moving a second point much. To be a little more definite,
let us say that a space X has the *Weak Effros Property* iff
for every pair $(p,q) \in X^2$ and every neighborhood $N(q)$ of

q there is a neighborhood N(p) of p such that for each
x ∈ N(p) there is a space homeomorphism h taking X onto
itself such that h(p) = x and h(q) ∈ N(q). (In other
words, we can jiggle p a little without moving q much.)

My first attempt at proving the Main Theorem resulted in
the following easier theorem about the Weak Effros Property.

THEOREM 12: A continuum which has the Weak Effros Property
and is separated by some pair of points must be a Hausdorff
circle.

The Main Theorem would be an easy corollary of Theorem 12
if only we had an affirmative answer to the following question.

Question 1. Does every homogeneous continuum have the
Weak Effros Property?

A well known theorem of Effros implies that every homogeneous *metric* continuum has the Weak Effros Property.

If we cannot show that the Main Theorem is a corollary of
Theorem 12, let us show that Theorem 12 is a corollary of the
Main Theorem:

Proof of Theorem 12. First show that a connected space
with the Weak Effros Property is homogeneous. Then apply the
Main Theorem. (Of course, my original proof did not involve
the use of the Main Theorem.)

Some other questions along these lines are:
Question 2. Does every homogeneous Hausdorff circle have
the Weak Effros Property?

Question 3.   Is every homogeneous Hausdorff circle
2-homogeneous?

Question 4.   Does every homogeneous Hausdorff circle admit
orientation reversing self homeomorphisms?

Question 5.   Is every homogeneous Hausdorff circle
3-homogeneous?

Question 6.   If a homogeneous continuum is separated by a
countable subset, must the continuum be a Hausdorff circle?

A Classification of Certain Inverse Limit Spaces

Will Watkins

What follows is an overview of my dissertation, subsequent results, questions and conjectures.

Let $I = [0,1]$. Let ^Nf be the N-th degree hat function from I to I. For example, ^2f, ^3f, and ^4f are pictured below:

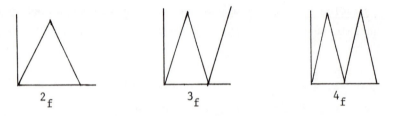

^2f ^3f ^4f

We are interested in classifying the spaces which are inverse limits of the unit interval using these bonding maps. In particular, for a fixed integer $N \geq 2$, we are interested in classifying (up to homeomorphism) the space D_N, which is $\varprojlim\{I, {}^Nf\}$. These spaces are often called Knaster type continua since D_2 is in fact the Knaster bucket handle.

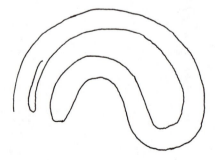

$\cong D_2$

Recently David Bellamy used D_6 to construct his celebrated example of a tree-like continua without the fixed point property. It appears improbable that his techniques can be modified to construct a similar example from D_2. This resurrects a question raised in a paper by J. W. Rogers, Jr.: Are there three topologically different D_N's? The answer is:

THEOREM: D_N is homeomorphic to D_M if and only if M and N have the same prime factors.

Assuming they have the same prime factors it is easy to demonstrate an inverse limit homeomorphism between the two.

The present objective will be to outline, without proofs, the steps in showing that D_2 is not homeomorphic to D_6. The proofs appear in my thesis and will appear, hopefully, elsewhere.

Consider the composant of D_2 and D_6 containing the endpoint as parametrized below.

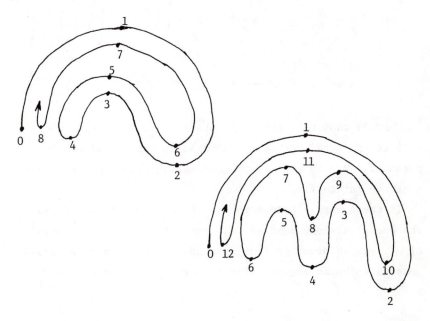

Consider the special basis about the point   0   in D_2 and D_6.

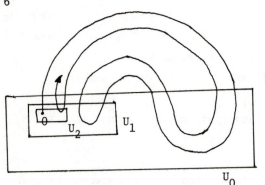

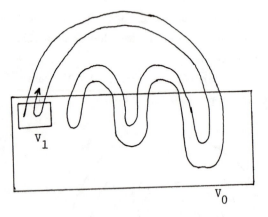

Observe that the integer points in U_i is exactly the collection $\{2n2^i : n$ is a non-negative integer$\}$.   The integer points in V_i is exactly the collection $\{2n6^i : n$ is a non-negative integer$\}$.

If there were a homeomorphism $h : D_2 \to D_6$ it would take the end-point-composant of D_2 onto the end-point-composant of D_6 in an order preserving manner.   Furthermore we could construct the following infinite lattice of open sets.

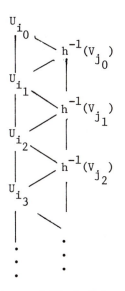

where $h(U_{i_0}) \subset V_0$.

We need two definitions. Suppose $A = \{a_i\}_{i=0}^{\infty}$ and
$B = \{b_i\}_{i=0}^{\infty}$ are two increasing sequences of non-negative
integers, then:

$A \underset{o}{\supseteq} B$ if and only if $a_0 = b_0 = 0$ and $b_i \in A$ for every i.

$A \underset{o}{\textcircled{k}} B$ if and only if $A \underset{o}{\supseteq} B$ and $b_i = a_{ki}$ for every i.

Now we construct, from our lattice of open sets, a lattice
of sequences. First we get the chain:

A_0

A_1

A_2

$\cdot$
$\cdot$
$\cdot$

where A_n is the collection of integers in U_{i_n}, and for every i there is an integer r_i so that $A_i \xrightarrow[0]{\overline{2^{r_i})}} A_{i+1}$. Each A_i is an arithmetic sequence. Unfortunately, the integers in V_{j_0} may not be mapped to integers in U_{i_0} under h^{-1}. However, each integer in V_{j_0} will be mapped into some arc component of U_{i_0} and at most one integer is mapped into any arc component. Thus, using a "nearest integer function," this allows us to construct a lattice:

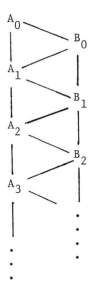

B_n is the subsequence of A_n obtained by picking those integers in U_{i_n} that are on the same arc component of U_{i_n} containing some $h^{-1}(m)$ where m is some integer point in V_{j_n}.

This lattice has the properties that for every i there exist r_i and s_i so that $A_i \xrightarrow[0]{\overline{2^{r_i})}} A_{i+1}$, $B_i \xrightarrow[0]{\overline{6^{s_i})}} B_{i+1}$ and $A_i \underset{0}{\supset} B_i$ and $B_i \underset{0}{\supset} A_{i+1}$.

At one time I had hoped to show that no such lattice exists. I have been unable to do this.

By picking some special but very natural chainings of D_2 and D_6 we can establish one more useful fact. We pick a nested sequence of chainings of $D_2 - A_0 \gg A_1 \gg A_2 \gg A_3 \gg \cdots$ where $A_i \gg A_{i+1}$ means A_{i+1} refines A_i. Further, U_i will be the first link in A_i. Similarly pick a nested sequence of chainings of $D_6 - B_0 \gg B_1 \gg B_2 \gg \cdots$.

We could then get an infinite lattice of chainings:

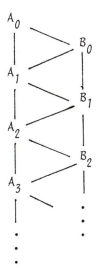

Use the first link in each chaining to construct the same lattice of sequences we had earlier. Knowledge about the chainings helps establish that B_1 must be an arithmetic sequence! (This is non-trivial.)

For any arithmetic sequence C let δC be the difference between two consecutive elements. Since B_1 is arithmetic and $B_1 \overset{s_1}{\underset{o}{\supseteq}} B_2$ we see $\delta B_2 = 6^{s_1} \delta B_1$. Since $B_2 \overset{s_1}{\underset{o}{\supseteq}} A_3$ there is some constant k so that $\delta A_3 = k \delta B_2 = k 6^{s_1} \delta B_1$. We know that δA_3 is some power of 2 and this is a contradiction.

We now have the foundation for showing that there are countably many Knaster type continua with fixed bonding maps. These techniques can be applied to a larger class of Knaster

continua. Let K and L be two distinct subsets of the
primes. Construct inverse limit spaces $D_{\widetilde{K}}$ and $D_{\widetilde{L}}$ where
the bonding maps are hat functions and each element of K is
a factor of the degree of infinitely many of the bonding maps
in $D_{\widetilde{K}}$ (similarly for $D_{\widetilde{L}}$). Then $D_{\widetilde{K}}$ and $D_{\widetilde{L}}$ are not
homeomorphic. This shows that there are precisely c of
these continua. I do not as yet know necessary conditions for
$D_{\widetilde{K}}$ and $D_{\widetilde{L}}$ to be homeomorphic.

These results became of interest to me as a consequence of
David Bellamy's construction. His example cannot be embedded
in the plane. The question arises: Can we use his techniques
to construct a planar tree-like continua without the fixed
point property? Possibly we can search through the Knaster
type continua and locate those possessing self homeomorphisms
with only one fixed point. (The fixed point will necessarily
be the 0 point.) Then investigate ways of embedding that
continuum in the plane. Suppose we found $h : D_{\widetilde{N}} \to D_{\widetilde{N}}$ and the
embedding

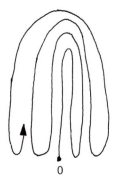

0

so that any sequence of points $\langle x_i \rangle$ converging to the 0
point from the left had the property that $\langle h(x_i) \rangle$ converged
to 0 from the right and any sequence $\langle y_i \rangle$ converging from
the right had the property that $\langle h(y_i) \rangle$ converged from the

left. Then, much as David Bellamy did, we could split the end
point. We only need to recompactify over a two point set
however.

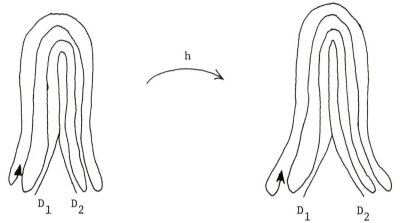

h

D_1 D_2 D_1 D_2

Define $\bar{h}(0_1) = \bar{h}(0_2)$ and finish constructing $\bar{h}$ from h in
the obvious manner. We then would have a planar tree-like
continuum without the fixed point property.

A possibly useful result completed since my thesis states:

THEOREM: If $h : D_{\widetilde{N}} \to D_{\widetilde{N}}$ is a homeomorphism with h(0) = 0
then there exists two inverse limit homeomorphisms $W_a : D_{\widetilde{N}} \to D_{\widetilde{N}}$
and $W_b : D_{\widetilde{N}} \to D_{\widetilde{N}}$ so that h is homotopic to $W_b^{-1} W_a : D_{\widetilde{N}} \to D_{\widetilde{N}}$.
This is proved by showing there exists a uniformly continu-
ous homotopy defined on the one-end-point-composant containing
0 (a dense subset). The proof that B_1 is arithmetic is very
useful here.

Finally a conjecture and a question.

Conjecture: If any element in the homotopy class of a self
homeomorphism of D_N has a fixed point on a given composant
then every element in that homotopy class has a fixed point on
that same composant.

<u>Question</u>:   Is there a homeomorphism $h : D_{\tilde{N}} \to D_{\tilde{N}}$ with one fixed point and an embedding of $D_{\tilde{N}}$ in the plane so that the image of any sequence converging from the right to 0 converges from the left and vice versa?

REFERENCES

Watkins, William Thomas, "Homeomorphic classification of inverse limit spaces with fixed open bonding maps," Ph.D. thesis, University of Wyoming, 1980.

Bellamy, David, "A tree-like continuum without the fixed point property," preprint.

Rogers, J. W., Jr., "On mapping indecomposable continua onto certain chainable indecomposable continua," <u>Proc. Amer. Math. Soc.</u> 25 (1969), 449–456.

A Note Concerning a Continuum of J. C. Mayer

Sam W. Young

1) Introduction. Elsewhere in these proceedings, [4] Mayer describes an example of a tree-like continuum which has some of the same properties as the continuum of Ingram [3] but which has an embedding in the plane with a simple dense canal. Mayer's continuum, which he calls the X-odic continuum, is constructed as the limit of an inverse limit system on X's (4-ods). This author raised the question as to whether or not the X-odic continuum is in fact a bit simpler by being simple-triod-like. The purpose of this note is to provide an affirmative answer to that question.

For more information and background see [1] and [2].

2) Our beginning point will be the schematic diagram in figure 1 of [4] which is reproduced here in diagram 3. Denote the 4-od ABCD by $+$ and the triod EFG by $\perp$. Denote the Mayer bonding map of diagram 3 by $f: + \to +$. Let $h: + \to \perp$ be as in diagram 1 and $g: \perp \to +$ be as in diagram 2. The Mayer continuum X is the limit of the inverse limit system $+ \xleftarrow{f} + \xleftarrow{f} + \xleftarrow{f} \cdots$.

THEOREM. The continuum X is simple-triod-like.

Proof. All that needs to be shown is that the mapping f factors through a triod. This is accomplished by $f = gh = g(h)$. We are

$$+ \xleftarrow{f} +$$

$$g \nwarrow \perp \nearrow h$$

following the same convention as [4] and so it is hoped that the schematic diagrams will be self explanatory. The mapping h "unwraps" the arcs $[0,C]$ and $[2A/3, A]$ and the mapping g "wraps" them back again.

So now X is the limit of the system $+ \frac{gh}{\leftarrow} + \frac{gh}{\leftarrow} + \frac{gh}{\leftarrow} \cdots$ which is equivalent to $\perp \frac{hg}{\leftarrow} \perp \frac{hg}{\leftarrow} \perp \frac{hg}{\leftarrow} \cdots$. Thus the continuum X is the limit of an inverse limit system on simple triods and therefore is simple-triod-like. □

The bonding map hg is represented schematically in diagram 4. Although the bonding map hg does not seem to help in establishing the main properties of the continuum X, it must inevitably detract from its name.

REFERENCES

[1]   Beverly Brechner and John C. Mayer, "The Prime End Structure of Indecomposable Continua and the Fixed Point Property," to appear in Proceedings of Conference on Topology and Set Theory, held at Riverside, Calif., May 1980.

[2]   Beverly Brechner and John C. Mayer, "The Prime End Structure of Indecomposable Continua and the Fixed Point Property," these proceedings.

[3]   W. T. Ingram, "An Atriodic Tree-like Continuum with Positive Span," Fund. Math. 77 (1972), 99-107.

[4]   John C. Mayer, "Principal Embeddings of Atriodic Plane Continua," these proceedings.

On the Relative Complexity of a Tree-like Continuum and Its Proper Subcontinua

Sam W. Young

1) Introduction. One of the remarkable features of the continuum of W. T. Ingram [5] is the "gap" in complexity between the continuum and its proper subcontinua. Specifically, the continuum is T-like (simple-triod-like), not arc-like and every proper subcontinuum is arc-like. This combination of structural properties leads us to ask if there is a continuum with an even wider "gap."

Question 1. Does there exist a continuum which is 4-od-like, not simple-triod-like and every proper subcontinuum is arc-like?

Definition. The tree-like continuum M is said to be *finitely junctioned* if and only if there exists a positive integer j such that M is Π_j-like where Π_j denotes the collection of all trees which have no more than j points of order >2 (junction points). A tree-like continuum which is not finitely junctioned is *infinitely junctioned*.
Pursuing an even larger "gap" we ask the following:

Question 2. Does there exist an infinitely junctioned tree-like continuum every proper subcontinuum of which is arc-like?
The main purpose of this paper is to give two examples of continua which are infinitely junctioned and every proper subcontinuum is simple-triod-like. The second example improves upon the first by having the additional property of

being atriodic. The reader should keep in mind that an arc-
like continuum is simple-triod-like but the converse is not
true.

In section 3) we discuss some properties of infinitely
junctioned continua. Relevant to Question 1 is [9] elsewhere
in these proceedings. For basic definitions and terminology,
see [7] and [5]. The terms "arc-like" and "chainable" are
used interchangeably depending upon context.

2)   The continuum A and the continuum B. A is obtained as
the limit of an inverse limit system $T_1 \xleftarrow{f_1} T_2 \xleftarrow{f_2} T_3 \xleftarrow{f_3} \cdots$
where for each i, T_i is a tree and f_i is a mapping of
T_{i+1} onto T_i. For $i \geq 1$, $T_i = [(\frac{2}{3},0), (i + \frac{1}{3},0)] \cup$
$\bigcup_{j=1}^{i} [(j,0),(j,1)]$. Let τ_j^i denote the simple triod
$[(j -\frac{1}{3}, 0), (j + \frac{1}{3},0)] \cup [(j,0),(j,1)]$ $1 \leq j \leq i$,
$i = 1, 2, 3, \cdots$ and α_j^i denotes the arc
$[(j +\frac{1}{3},0), (j + \frac{2}{3},0)]$ $1 \leq j \leq i -1$, $i = 2, 3, 4, \cdots$.
For $i \geq 1$, the bonding map $f_i : T_{i+1} \to T_i$ is described in
three parts: 1) For $1 \leq j \leq i$, f_i fixes each point of
τ_j^{i+1}, 2) f_i takes each point of τ_{i+1}^{i+1} to $(i + \frac{1}{3},0)$ and
3) f_i is extended to a mapping of T_{i+1} onto T_i in such
a way that each of the arcs α_j^{i+1} maps onto T_i $1 \leq j \leq i$.
See illustration 1 for a picture of $f_2 : T_3 \to T_2$.

Note that the extension used in illustration 1 is not the
most efficient to accomplish part 3). But, in general, if we
pinch each of the arcs α_j^{i+1} and spiral counterclockwise,
then the entire picture of f_i can be drawn in the plane.
Therefore we claim 1) of the following:

THEOREM 1:   The continuum A has the following properties:
1) A is planar, 2) A contains an infinite collection of

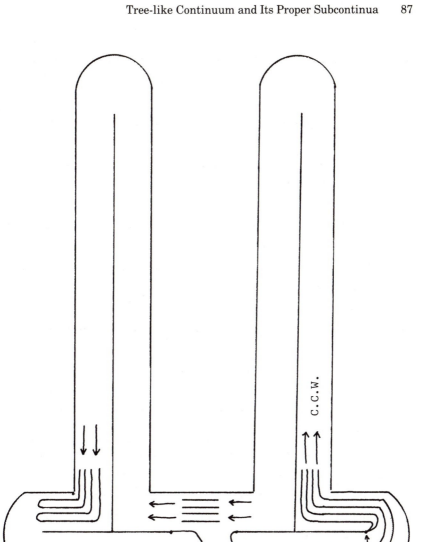

C.C.W.

X = 1   X = 2 $\tau \, {}_3^3$

Fig. 1.   The continuum A. $f_2 : T_3 \to T_2$

mutually exclusive simple triods, and 3) Every proper sub-continuum of A is simple-triod-like.

Proof of Property 2. If $j \geq 1$ and each of the bonding maps f_j, f_{j+1}, f_{j+2}, $\cdots$ is restricted to τ_j^{j+1}, τ_j^{j+2}, τ_j^{j+3} $\cdots$ respectively, the result is a proper subcontinuum of A which is a simple triod.

Proof of Property 3. Suppose K is a proper subcontinuum of A and $\varepsilon > 0$. There exists an integer n such that $\Pi_n : A \rightarrow T_n$, the nth projection map, is an ε-map and $\Pi_n(K)$ is a proper subcontinuum of T_n. Now $\Pi_{n+1}(K)$ does not contain any of the arcs α_j^{n+1} $j \leq n$ since each of them is mapped onto T_n by f_n. The map Π_{n+1} is an ε-map of K onto $\Pi_{n+1}(K)$ and $\Pi_{n+1}(K)$ must be a simple triod. So by definition, K is simple-triod-like.

The construction of the continuum B will resemble that of A to some extent. The same trees T_i will be used as the factor spaces but the bonding maps f_i will contain copies of the Ingram mapping [5]. We will not reproduce the detailed description of the Ingram mapping here but we will consistently employ his notation for the end points. Let a, b, c and 0 denote respectively the end points and the junction point of a simple triod τ. If $f : \tau \rightarrow \tau$ is the Ingram mapping, then c will be the end point for which $f(a) = f(b) = f(c) = c$. The end point $b = f(0)$.

Now let $a_j^i = (j - \frac{1}{3}, 0)$, $b_j^i = (j,1)$, $c_j^i = (j + \frac{1}{3}, 0)$ and $0_j^i = (j,0)$ for $1 \leq j \leq i$, i = 1, 2, 3, $\cdots$. Thus a_j^i, b_j^i, c_j^i and 0_j^i are, respectively, the left end point, upper end point, right end point and junction point of the simple triod τ_j^i.

The bonding map $f_i : T_{i+1} \to T_i$ is defined in three parts:
1) For $1 \leq j \leq i$, f_i maps τ_j^{i+1} onto τ_j^{i+1} and is a
copy of the Ingram mapping such that $f_i(a_j^{i+1}) = f_i(b_j^{i+1}) =$
$= f(c_j^{i+1}) = c_j^i$ and $f_i(O_j^{i+1}) = b_j^i$. 2) f_i takes each point
of τ_{i+1}^{i+1} to $(i + \frac{1}{3}, 0) = c_i^i$, and 3) f_i is extended to a
mapping of T_{i+1} onto T_i in such a way that each of the
arcs α_j^{i+1} maps onto T_i, $1 \leq j \leq i$.

See illustration 2 for a picture of $f_2 : T_3 \to T_2$. As before we can make each of the arcs $\alpha_j^{i+1} = [c_j^{i+1}, a_j^{i+1}]$ map
onto T_i by pinching and spiraling counterclockwise. So we
claim 1) of Theorem 2.

THEOREM 2: The continuum B has the following properties:
1) B is planar, 2) B contains an infinite collection of
mutually exclusive non-chainable subcontinua (Ingram continua),
3) Every proper subcontinuum of B is simple-triod-like, and
4) B is atriodic.

The proofs of properties 2) and 3) follow those of Theorem
1. In order to prove property 4), we must have the following
definition and lemma:

Definition: If each of T_1 and T_2 is a tree and f is
a mapping of T_1 onto T_2 then f is said to have the
three point property if and only if it is true that given any
three points of T_1, the image of one of them lies in the
image of the arc which joins the other two.

LEMMA: (the three point lemma) If the continuum M is
the limit of the inverse limit system $T_1 \xleftarrow{f_1} T_2 \xleftarrow{f_2} T_3 \xleftarrow{f_3} \cdots$
and infinitely many of the bonding maps have the three point
property, then M is atriodic.

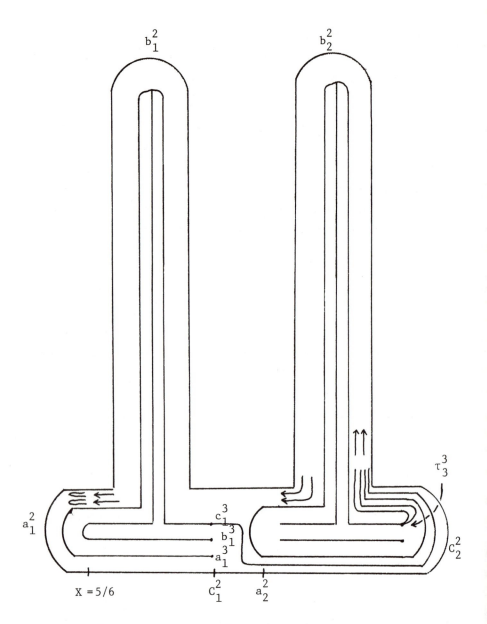

Fig. 2.   The continuum B. $f_2 : T_3 \to T_2$

Proof. Suppose to the contrary that M contains a triod K. Then K is the proper union of three continua K_1, K_2, and K_3 such that the intersection of each two of them is the intersection of all three. Let X_1, X_2 and X_3 be points of $K_1 \sim (K_2 \cup K_3)$, $K_2 \sim (K_1 \cup K_3)$ and $K_3 \sim (K_1 \cup K_2)$ respectively. There exists a positive integer n such that f_n has the three point property and for $i \geq n$, $\Pi_i(X_1) \notin \Pi_i(K_2 \cup K_3)$, $\Pi_i(X_2) \notin \Pi_i(K_1 \cup K_3)$ and $\Pi_i(X_3) \notin \Pi_i(K_1 \cup K_2)$. Consider the points $\Pi_{n+1}(X_1)$, $\Pi_{n+1}(X_2)$ and $\Pi_{n+1}(X_3)$ of T_{n+1}. For one of them, $\Pi_{n+1}(X_1)$ say, $f_n(\Pi_{n+1}(X_1)) = \Pi_n(X_1)$ lies in the image under f_n of the arc α joining $\Pi_{n+1}(X_2)$ to $\Pi_{n+1}(X_3)$. But $f_n(\alpha) \subset \Pi_n(K_2 \cup K_3)$ and we have a contradiction.

Proof of Property 4. We will show that each of the bonding maps has the three point property. Let i be a positive integer and let X_1, X_2 and X_3 be three points of T_{i+1}. Note that if two of the points, X_1 and X_2 say, are separated by one of the arcs α_j^i then since $f_i(\alpha_j^i) = T_i$, then of course $f(X_3)$ lies in the image of the arc joining X_1 to X_2. If no two points are separated by one of the arcs α_j^i, then they lie together on $\tau_1^{i+1} \cup \alpha_1^{i+1}$, the "first" triod, or on $\tau_{i+1}^{i+1} \cup \alpha_i^{i+1}$, the "last" triod, or on $\tau_j^{i+1} \cup \alpha_{j-1}^{i+1} \cup \alpha_j^{i+1}$, a "middle" triod in case $i \geq 2$ and $1 < j < i$. Since all of the bonding maps are essentially the same except for having more "middle" triods, we can handle all three cases by just showing that $f_2 : T_3 \to T_2$ has the three point property. See illustration 2 again.

Any arrangement for which X_1, X_2 and X_3 lie together on an arc is a trivial case and so we proceed with:

Case "last." Suppose X_1, X_2 and $X_3 \in \tau_3^3 \cup \alpha_2^3$ and $X_1 \in [(3,0), (3.k)]$. Then another point, X_2 say, lies in $[(2 + \frac{1}{3}, 0), (3,0)]$ and $X_3 \in [(3,0), (3 + \frac{1}{3}, 0)]$. Since f_2 takes all of τ_3^3 to a point, it follows that $f(X_1) = f(X_3)$.

Case "first." Suppose $X_1 \in [(\frac{2}{3}, 0), (1,0)]$, $X_2 \in [(1,0), (1,1)]$ and $X_3 \in [(1,0), (1 + \frac{2}{3}, 0)]$. If $X_3 \in [(1,0), (1 + \frac{1}{3}, 0)]$, then X_1, X_2 and X_3 all lie in τ_1^3 and f_2 restricted to τ_1^3 is an Ingram mapping. A short analyses of the Ingram mapping will show that it has the three point property. We will not include that argument here but continue instead with the case $X_3 \in [(1 + \frac{1}{3}, 0), (1 + \frac{2}{3}, 0)]$. The image under f_2 of the arc joining X_3 to X_1 must include $[(1,0), (1 + \frac{1}{3}, 0)] \cup \cup [(1,0), (1,1)]$. Therefore $f_2(X_2)$ must lie in $[(\frac{2}{3}, 0), (1,0)]$. This leaves no possible position for X_3 without either $f(X_3)$ lying in the image of the arc joining X_3 to X_1 or $f(X_2)$ lying in the image of the arc joining X_3 to X_1.

Case "middle." Suppose $X_1 \in [(1 + \frac{1}{3}, 0), (2,0)]$, $X_2 \in [(2,0), (2,1)]$ and $X_3 \in [(2,0), (2 + \frac{2}{3}, 0)]$. In case $X_1 \in [(1 + \frac{2}{3}, 0), (2,0)]$, then our problem is reduced to case "first." So suppose $X_1 \in [(1 + \frac{1}{3}, 0), (1 + \frac{2}{3}, 0)] = \alpha_1^3$. The image under f_2 of the arc joining X_1 to X_3 must include $[(2 - \frac{1}{3}, 0), (2,0)] \cup [(2,0), (2,1)]$. Therefore $f(X_2)$ must lie on $[(2,0), (2 + \frac{1}{3}, 0)]$. As before, this leaves no possible position for X_3.

It has now been shown that each bonding map has the three point property and so by the three point lemma, the continuum B is atriodic.

3) In this section we make some observations about tree-like continua which are infinitely junctioned. We will make use of:

THEOREM 3: (H. Cook [2]) Suppose j is a positive integer and Π is a collection of finite graphs no one of which contains more than j points of order >2. Then if M is a Π-like continuum, there exists a finite subset V of j points of M such that every subcontinuum of M $\sim$ V is chainable.

Our Theorem 3 is Theorem 4 of [2] stated in its full generality.

THEOREM 4: Each of the continua A and B are infinitely junctioned.

Proof. Each of A and B contains an infinite collection of mutually exclusive non-chainable continua. Such a property cannot be possessed by a finitely junctioned continuum because of Theorem 3.

Definition. A tree-like continuum is said to be *hereditarily infinitely junctioned* if and only if every nondegenerate subcontinuum of it is infinitely junctioned.

THEOREM 5: A tree-like continuum is hereditarily infinitely junctioned if and only if it does not contain a nondegenerate chainable subcontinuum.

Proof. If the tree-like continuum M is hereditarily infinitely junctioned, then it follows immediately from the definition that M does not contain a nondegenerate chainable

continuum. Now suppose that M is a tree-like continuum
which does not contain a nondegenerate chainable subcontinuum.
Then M does not contain a finitely junctioned subcontinuum
either since Theorem 3 implies that any finitely junctioned
continuum must contain a nondegenerate chainable subcontinuum.

In [6] an example is given of an hereditarily indecomposable tree-like continuum containing only degenerate chainable
subcontinua. Theorem 5 implies that this continuum is hereditarily infinitely junctioned. So also would be the continuum
in [8] which is tree-like and contains only degenerate continuous images of chainable continua.

Definition. A continuum is said to be *hereditarily equiva-*
lent if and only if it is homeomorphic to each of its nondegenerate subcontinua.

The only known examples of hereditarily equivalent continua
are the arc and the pseudo-arc. We end this section with a
theorem which describes the state of the art in search of a
third example.

THEOREM 6: An hereditarily equivalent continuum is either
an arc or a pseudo-arc or is an hereditarily infinitely junctioned tree-like continuum.

Proof. An hereditarily equivalent continuum must be tree-
like [3]. If it is chainable then it is either the arc or
the pseudo-arc depending upon whether or not it is decomposable
[4], [1]. If it is not chainable then it does not contain a
nondegenerate chainable subcontinuum. So by Theorem 5, it is
hereditarily infinitely junctioned.

REFERENCES

[1] Bing, R. H., "Concerning hereditarily indecomposable continua." Pacific Journal of Math. 1 (1951), 43-51.

[2] Cook, H, "Concerning three questions of Burgess about homogeneous continua." Colloq. Math. 19 (1968), 241-244.

[3] ------, "Tree-likeness of hereditarily equivalent continua." Fund. Math. 68 (1970), 203-205.

[4] Henderson, G. W., "Proof that every compact decomposable continuum which is topologically equivalent to each of its nondegenerate subcontinua is an arc." Ann. Math. 27 (1960), 421-428.

[5] Ingram, W. T., "An atriodic tree-like continuum with positive span." Fund. Math. 77 (1972), 99-107.

[6] ------, "Hereditarily indecomposable tree-like continua, II." Fund. Math. (to appear).

[7] Mardesić, S. and Segal, J. "ε-mappings onto polyhedra." Trans. A.M.S. 109 (1963), 146-164.

[8] Rogers, J. W., "Continua that contain only degenerate continuous images of plane continua." Duke Math. Journal 38 (1970), 479-483.

[9] Young, S. W., "A note concerning a continuum of J. C. Mayer." These proceedings.

Saturated 2-Sphere Boundaries in Bing's Straight-Line Segment Example

Steve Armentrout

1. INTRODUCTION

In [1], Woodruff proved that if G is an upper semicontinuous decomposition of S^3 such that each point of S^3/G has arbitrarily small neighborhoods bounded by 2-spheres missing the image of the nondegenerate elements of G, then S^3/G is homeomorphic to S^3. In [6], Woodruff gives a number of extensions of the result above. She also raised the following question: Suppose G is an upper semicontinuous decomposition of S^3 such that each element of G has arbitrarily close neighborhoods bounded by 2-spheres saturated relative to G. Is S^3/G then homeomorphic to S^3?

In this note, we shall show that Bing's straight-line segment example [2], regarded as a decomposition of S^3, can be used to answer the question of Woodruff's stated above. We shall prove that each element of Bing's decomposition has arbitrarily close neighborhoods bounded by saturated 2-spheres. Eaton's results [4] imply that for this decomposition G of S^3, S^3/G is not homeomorphic to S^3. Thus Woodruff's question has a negative answer.

2. SOME NOTATION

We shall follow Bing's description and notation [2]. The decomposition of section 6 of [2] is a decomposition H of E^3.

In his construction, Bing uses two horizontal planes Θ_1 and Θ_2 with Θ_1 below Θ_2. He describes a sequence $G_1, G_2, G_3, \cdots$ of finite graphs in E^3 and, for each n, an open tubular neighborhood U_n of G_n. These neighborhoods have the property that for each n, $\overline{U}_{n+1} \subset U_n$.

Let A_1, A_2, A_3, $\cdots$ and B_1, B_2, B_3, $\cdots$ be two sequences of horizontal planes between Θ_1 and Θ_2 such that the following hold: (1) A_1 is above B_1. (2) For each n, A_{n+1} is above A_n. (3) The sequence A_1, A_2, A_3, $\cdots$ converges to Θ_2. (4) For each n, B_{n+1} is below B_n. (5) The sequence B_1, B_2, B_3, $\cdots$ converges to Θ_1. (6) For each n, the closure of each component of $U_n - (A_n \cup B_n)$ lying between A_n and B_n is a nearly vertical cylinder which intersects only one component of $G_n - (A_n \cup B_n)$.

Suppose X and Y are two of A_1, A_2, $\cdots$, B_1, B_2, $\cdots$, and for some n, C is a nearly vertical cylinder which is the closure of a component of $U_n - (X \cup Y)$. We shall use the terms *top*, *base*, and *side* of C in an obvious way.

Now we shall describe a notation for components of the sets U_1, U_2, U_3, $\cdots$. U_1 is connected. U_2 has eight components and we shall denote them by U_{11}, U_{12}, $\cdots$, and U_{18}. If $1 \leq i \leq 8$, there are eight components of $U_{1i} \cap U_3$, and we shall denote them by U_{1i1}, U_{1i2}, $\cdots$, and U_{1i8}. Continue this process. If n is any positive integer greater than one and U is a component of U_n, there is a finite sequence i_2, i_3, $\cdots$, and i_n of integers such that (1) if $1 < j \leq n$, $i_j = 1, 2, \cdots$, or 8, and (2) U is denoted by $U_{1i_2 i_3 \cdots i_n}$.

3. A BASIC 2-SPHERE

Suppose g is a nondegenerate element of H and ϵ is a positive number. We shall show how to obtain a particular 2-sphere S_g in E^3 such that $g \subset \text{int } S_g$, $S_g \cup \text{int } S_g$ lies in the ϵ-neighborhood of g, and each element of H that intersects S_g intersects it in only one point. We regard the 2-sphere we shall construct as a "basic" 2-sphere about g, and we shall modify S_g to obtain a saturated 2-sphere Σ_g which bounds a neighborhood of g that lies in the ε-neighborhood of g. The construction of such a basic 2-sphere was described (for a slightly different decomposition) in [1].

Let i_2, i_3, i_4, $\cdots$ be the sequence of indices such that for each n, $g \subset U_{1i_2i_3\cdots i_n}$. Then $g = \cap_{n=2}^{\infty} U_{1i_2i_3\cdots i_n}$, and it follows that for some m, $\overline{U}_{1i_2i_3\cdots i_m}$ lies in the ε-neighborhood of g. Let α denote the index $1i_2i_3\cdots i_m$.

$\overline{U}_\alpha - (A_1 \cup B_1)$ has exactly 14 components, two above A_1, eight between A_1 and B_1, and four below B_1. Of these, only three intersect g, one above A_1, one between A_1 and B_1, and one below B_1. Let W_g be the closure of the union of those three components of $\overline{U}_\alpha - (A_1 \cup B_1)$. W_g is a 3-cell and let S_g = Bd W_g. S_g is a *basic 2-sphere about* g.

$S_g \cap A_1$ is the union of a simple closed curve, and three mutually disjoint discs which we shall denote by D_2, D_3, and D_4. $S_g \cap B_1$ is the union of a simple closed curve and a disc, and we shall denote this disc by D_1. See Figure 1.

We shall describe how to modify the D's so as to obtain discs Δ_1, Δ_2, Δ_3, and Δ_4 such that (1) if i = 1, 2, 3, or 4, D_i and Δ_i have the same boundary and (2) if we replace the D's by the Δ's, we obtain a saturated 2-sphere Σ_g which bounds a neighborhood of g lying in the ε-neighborhood of g.

We shall describe the construction of Δ_1. This construction may be easily modified to obtain Δ_2, Δ_3, and Δ_4.

4. MODIFICATION OF D_1

Recall that W_g is constructed using the component U_α of U_m. There is a unique component of U_{m+1} that intersects

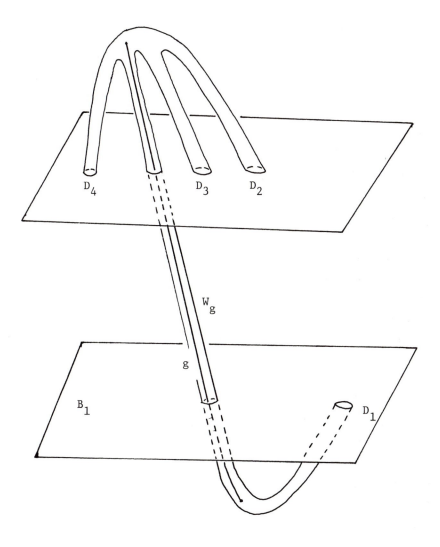

Figure 1.

D_1;   let   r   be the index such that this component of U_{m+1}
is $U_{\alpha r}$.

    If $p > m$ and   V   is a component of U_p intersecting D_1,
each component of $\overline{V} \cap D_1$ is a disc.   Let M_1, M_2,$\cdots$,   and
M_8 denote the components of $D_1 \cap \overline{U}_{\alpha r}$.   If $i = 1, 2, \cdots$,
or   8,   let M_{i1}, M_{i2},$\cdots$,   and M_{i8} denote the components
of $M_i \cap \overline{U}_{m+2}$.   Continue this notational scheme.

    There exists a spanning arc α_1 of D_1 such that if   q
is any positive integer and each of i_1, i_2,$\cdots$,   and i_q is
either   1, 2,$\cdots$,   or   8,   then $\alpha_1 \cap M_{i_1 i_2 \cdots i_q}$ is a spanning arc of $M_{i_1 i_2 \cdots i_q}$.   The arc α_1 divides D_1 into two
subdiscs D_1^+ and D_1^-.

    Let C_1, C_2,$\cdots$,   and C_8 denote the closures of the com-
ponents of $U_{\alpha r} - (A_1 \cup B_1)$ lying between A_1 and B_1,
indexed so that if $1 \leq i \leq 8$, $M_i = C_i \cap D_1$.   Let
N_1, N_2,$\cdots$,   and N_8 be mutually disjoint discs in   Int D_{11}
such that if $i = 1, 2, \cdots$,   or   8, $N_i \cap M_i$ is an arc β_i
on   Bd M_i and N_i intersects no one of M_1, M_2,$\cdots$,   and
M_8 other than M_i.

    Suppose $i = 1, 2, \cdots$,   or   8.   Let F_i be a tubular fin-
ger with base N_i and which extends upwards along the side of
C_i as indicated in   Figure 2. F_i is a 3-cell, and also is
a nearly vertical cylinder with base N_i and top on A_1.
Further, $F_i \cap C_i$ is a disc d_i which we may think of as a
vertical strip running from A_i to B_i.

    If $i = 1, 2, \cdots$,   or   8,   let Q_{i1}, Q_{i2},$\cdots$,   and Q_{i8}
denote the components of $\overline{U}_{m+2} \cap C_i$,   indexed so that if
$1 \leq j \leq 8$, $Q_{ij} \cap D_1 = M_{ij}$.   Now $\alpha_1 \cap M_i$ is a spanning arc
of M_i,   and divides M_i into two subdiscs, $M_i^+ \subset D_1^+$ and
$M_i^- \subset D_1^-$.   Note that β_i lies on   Bd M_i^+.

    Let λ_{i1}, λ_{i2},$\cdots$,   and λ_{i8} be mutually disjoint arcs in
M_i^+,   each with one endpoint on   Int β_i and such that if

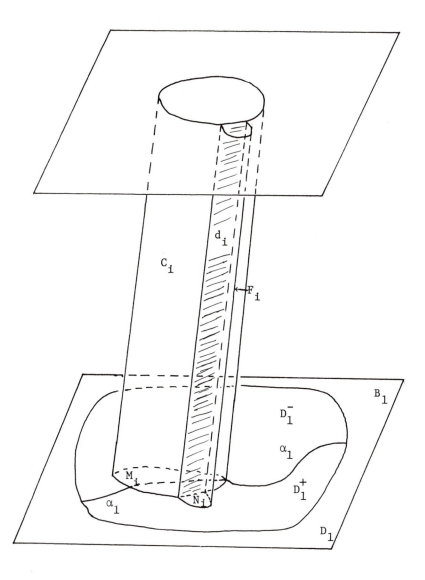

Figure 2.

$j = 1, 2, \cdots$, or 8, λ_{ij} has an endpoint on Bd M_{ij} and Int λ_{ij} is disjoint from M_{i1}, $M_{i2}, \cdots$, and M_{i8}. See Figure 3.

We shall use the following lemma which is a corollary of Proposition 2.2 of [3].

LEMMA: Suppose C is a vertical cylinder in E^3 and σ_1, $\sigma_2, \cdots$, and σ_t are mutually disjoint straight-line segments in C such that if $1 \leq s \leq t$, σ_s has one end on the base of C and the other end on the top of C. Then there is a homeomorphism $h : C \to C$ such that (1) h is the identity on the base and side of C, (2) h is invariant on each horizontal disc of C, and (3) if $1 \leq s \leq t$, $h(\sigma_s)$ is vertical.

With the aid of the Lemma, it can be seen that if $1 \leq i \leq 8$, there is a homeomorphism $h_i : C_i \to C_i$ such that (1) h_i is the identity on the base and side of C_i, (2) h_i is invariant on horizontal discs of C_i, and (3) $h_i(Q_{i1})$, $h_i(Q_{i2}), \cdots$, and $h_i(Q_{i8})$ are parallel nearly vertical cylinders.

Suppose $i = 1, 2, \cdots$, or 8. If $1 \leq j \leq 8$, we may extend λ_{ij} to obtain a nearly vertical fin Λ_{ij} in C_i. Λ_{ij} is a disc whose boundary is the union of four arcs, (1) λ_{ij}, (2) an arc spanning d_i, (3) an arc lying in $Q_i \cap A_1$, and (4) an arc on the side of $h_i(Q_{ij})$. See Figure 4; there only one Q_{ij} is shown. The discs Λ_{i1}, $\Lambda_{i2}, \cdots$, and Λ_{i8} are to be mutually disjoint.

Thicken the discs Λ_{i1}, $\Lambda_{i2}, \cdots$, and Λ_{i8} slightly to obtain 3-cells Λ_{i1}^*, $\Lambda_{i2}^*, \cdots$, and Λ_{i8}^*. If $1 \leq j \leq 8$, let $L_{ij} = h_i^{-1}(\Lambda_{ij}^*)$. See Figure 5; there only one Q_{ij} is shown. For each j, $1 \leq j \leq 8$, let w_{ij} be the disc $L_{ij} \cap Q_{ij}$;

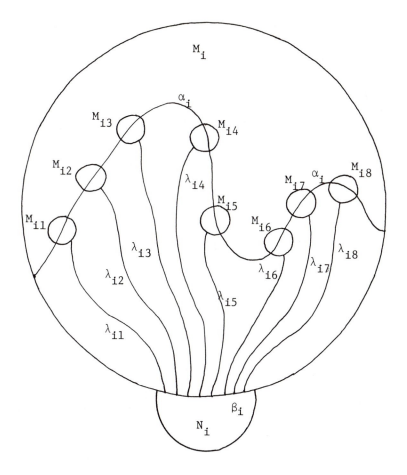

Figure 3.

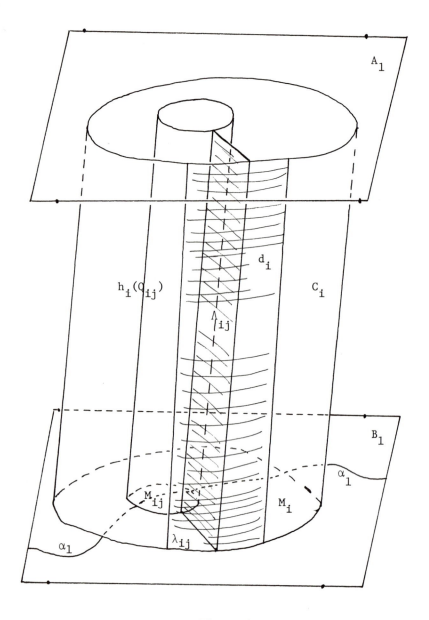

Figure 4.

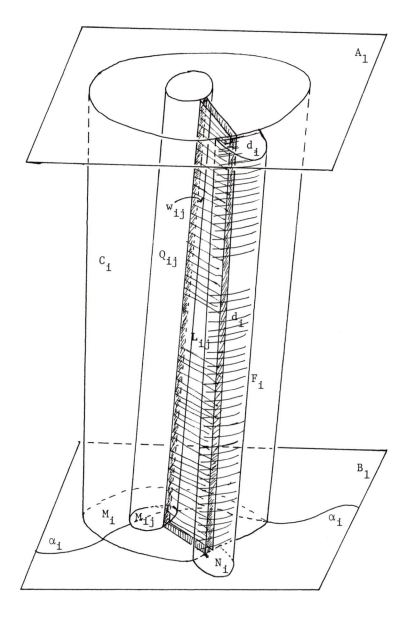

Figure 5.

w_{ij} is a thin nearly vertical strip that runs up the side of Q_{ij}.

Let P_{i1}, P_{i2}, $\cdots$, and P_{i8} be the closures of the components of $U_{m+2} - (A_1 \cup A_2)$ that lie between A_1 and A_2 and whose closures intersect C_i. We assume the notation such that if $1 \leq j \leq 8$, P_{ij} intersects Q_{ij}.

If $1 \leq j \leq 8$, let K_{ij} be a thin nearly vertical cylinder attached to the side of P_{ij} as indicated in Figure 6. $K_{ij} \cap P_{ij}$ is a disc z_{ij} such that $w_{ij} \cup z_{ij}$ is a disc d_{ij}. Further, K_{ij} intersects L_{ij} in a disc, so that $L_{ij} \cup K_{ij}$ is a 3-cell F_{ij}.

If $j = 1, 2, \cdots$, or 8, let $C_{ij} = P_{ij} \cup Q_{ij}$, let $N_{ij} = F_{ij} \cap A_1$, and let $\beta_{ij} = M_{ij} \cap N_{ij}$. Then we consider M_{ij1}, M_{ij2}, $\cdots$, M_{ij8} and repeat, for each i and j, $1 \leq i \leq 8$, $1 \leq j \leq 8$, the construction above, with C_{ij}, M_{ij}, the M_{ijk}, β_{ij}, and d_{ij} analogous to C_i, M_i, the M_{ij}, β_i, and d_i. First we construct fins, then thicken them, and then extend them upwards, in this case to the region between A_2 and A_3. This yields for each i, j, and k, a 3-cell F_{ijk}.

Continue this process. If p is any positive integer and i_1, i_2, $\cdots$, and i_p are integers, each equal to $1, 2, \cdots$, or 8, there is a 3-cell $F_{i_1 i_2 \cdots i_p}$ constructed by this process.

Suppose $i = 1, 2, \cdots$, or 8. For each n, let
$$X_{1i}^n = F_i \cup (\cup_{j=1}^8 F_{ij}) \cup \cdots \cup (\cup_{i_2=1}^8 \cup_{i_3=1}^8 \cdots \cup_{i_n=1}^8 F_{i i_2 i_3 \cdots i_n}).$$
For each n, X_{1i}^n is a 3-cell and $X_{1i}^n \subset X_{1i}^{n+1}$. Let $X_{1i} = C\ell[\cup_{n=1}^\infty X_{1i}^n]$. Then X_{1i} is a 3-cell. Further, X_{11}, X_{12}, $\cdots$, and X_{18} are mutually disjoint. Note that $X_{1i} \cap B_1$ is a disc Δ_{1i}^+ lying in D_1^+. Further, the discs Δ_{11}^+, Δ_{12}^+, $\cdots$, and Δ_{18}^+ are mutually disjoint.

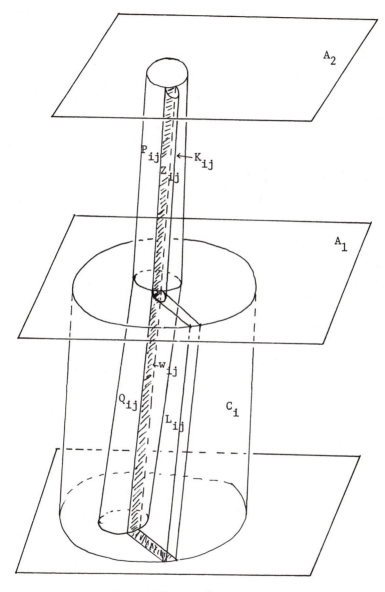

Figure 6.

It can be seen that if ℓ is any nondegenerate element of H intersecting D_1, there is an integer i, $1 \leq i \leq 8$, such that ℓ intersects M_i and if ℓ^+ is the part of ℓ on and above B_1, then $\ell^+ \subset \text{Bd } X_{1i}$.

By using an analogous procedure but starting in D_1^- and working downwards, we may construct mutually disjoint 3-cells $Y_{11}, Y_{12}, \cdots,$ and Y_{18} such that if $1 \leq i \leq 8$, $Y_{1i} \cap B_1$ is a disc Δ_{1i}^- lying in D_1^-. It follows that $\Delta_{11}^-, \Delta_{12}^-, \cdots,$ and Δ_{18}^- are mutually disjoint.

If ℓ is any nondegenerate element of H intersecting D_1, then for some i, $1 \leq i \leq 8$, ℓ intersects M_i, and if ℓ^- is the part of ℓ on and below B_1, then $\ell^- \subset \text{Bd } Y_{1i}$.

It is clear that if $1 \leq i \leq 8$, $X_{1i} \cap g = \phi$. For each i, $1 \leq i \leq 8$, Y_{1i} lies in the component of $\overline{U}_{m+1}$ intersecting D_1. This component of $\overline{U}_{m+1}$ does not intersect g. Hence if $1 \leq i \leq 8$, $Y_{1i} \cap g = \phi$.

If $1 \leq i \leq 8$, let $\hat{\Delta}_{1i}^+$ denote $(\text{Bd } X_{1i}) - (\text{Int } \Delta_{1i}^+)$, and let $\hat{\Delta}_{1i}^-$ denote $(\text{Bd } Y_{1i}) - (\text{Int } \Delta_{1i}^-)$. Let $\Delta_1^+ = \cup_{i=1}^8 \Delta_{1i}^+$, $\Delta_1^- = \cup_{i=1}^8 \Delta_{1i}^-$, $\hat{\Delta}_1^+ = \cup_{i=1}^8 \hat{\Delta}_{1i}^+$, and $\hat{\Delta}_1^- = \cup_{i=1}^8 \hat{\Delta}_{1i}^-$. Then let $\Delta_1 = [D_1 - \text{Int}(\Delta_1^+ \cup \Delta_1^-)] \cup (\hat{\Delta}_1^+ \cup \hat{\Delta}_1^-)$. Δ_1 is a disc, $\text{Bd } \Delta_1 = \text{Bd } D_1$, and $\Delta_1 \cap (S_g - \text{Int } D_1) = \text{Bd } D_1$. Further, $\Delta_1 \cap g = \phi$.

Roughly, the X_{1i}'s are feelers extending upwards and attached to the 3-cell W_g. The Y_{1i}'s are holes dug into W_g.

5. CONSTRUCTION OF Σ_g

By a procedure analogous to that used in section 4, we may construct, for $1 \leq i \leq 8$, 3-cells X_{2i}, Y_{2i}, X_{3i}, Y_{3i}, X_{4i}, and Y_{4i}, but, in these cases, with these X's and Y's attached to A_1 rather than B_1, and with the X's extending downwards and the Y's extending upwards. The X's are feelers attached to W_g, and the Y's are holes dug (upwards) into W_g.

As in section 4, we may, for $j = 2, 3$, or 4, construct a disc Δ_j such that Bd Δ_j = Bd D_j, $\Delta_j \cap (S_g - \text{Int } D_j)$ = Bd D_j, $\Delta_j \cap g = \phi$, and Δ_j lies in the component of $\bar{U}_{m+1}$ intersecting D_j.

Let Σ_g denote $(S_g - \cup_{i=1}^{4} \text{Int } D_i) \cup (\cup_{i=1}^{4} \Delta_i)$. Then Σ_g is a 2-sphere, $g \subset \text{Int } \Sigma_g$, Σ_g lies in the component of $\bar{U}_m$ containing g, and Σ_g is saturated relative to H.

We regard S^3 as the one-point compactification of E^3, and extend the decomposition H of E^3 to a decomposition G of S^3. It follows from [4] that S^3/G is not homeomorphic to S^3. Thus we have established the following result.

THEOREM: There exists a cellular decomposition G of S^3 such that (1) S^3/G is not homeomorphic to S^3 but (2) if g is any element of G, g has arbitrarily close neighborhoods in S^3 bounded by 2-spheres saturated relative to G.

ACKNOWLEDGMENT

This research was partly supported by National Science Foundation Grant MCS 78-11518.

REFERENCES

1. Armentrout, S., "A three-dimensional spheroidal space which is not a sphere." Fund. Math. 68 (1970), 183-186.

2. Bing, R. H., "Point-like decompositions of E^3." Fund. Math. 50 (1962), 431-453.

3. Bing, R. H. and Klee, V. L., "Every simple closed curve in E^3 is unknotted in E^4." J. London Math. Soc. 39 (1964), 86-94.

4. Eaton, W. T., "Applications of a mismatch theorem to decomposition spaces." Fund. Math. 89 (1975), 199-224.

5. Woodruff, E. P., "Decomposition spaces having arbitrarily small neighborhoods with 2-sphere boundaries." Trans. Amer. Math. Soc. 232 (1977), 195-204.

6. ------, "Decomposition spaces having arbitrarily small neighborhoods with 2-sphere boundaries II." Illinois J. Math. (to appear).

Decompositions of S³ into Circles

D. S. Coram

1. Introduction

This article presents some applications of work done by Coram
and Duvall [CD$_1$], [CD$_3$] to a question in decomposition space
theory. Namely, let G be an upper semicontinuous decomposition of S³ in which every element has the shape of a circle.
When is S³/G homeomorphic to S²? Perhaps the answer is
always, but additional conditions are imposed to get our results here.

THEOREM 1: Let G be an upper semicontinuous decomposition of S³ in which every element has the shape of a circle.
If i) for each g ∈ G there are saturated neighborhoods Ũ
and Ṽ of g in S³ such that for each decomposition element g' ⊂ Ṽ the inclusion map of g' into Ũ is essential,
and ii) S³/G is an ANR, then S³/G is homeomorphic to S²
and the decomposition map f : S³ → S³/G can be approximated
by Seifert fiber maps.

THEOREM 2: Let G be an upper semicontinuous decomposition of S³ in which every element has the shape of a circle.
If i) the decomposition map f : S³ → S³/G has the approximate homotopy lifting property, and ii) S³/G is finite
dimensional, then S³/G is homeomorphic to S² and f can
be approximated by Hopf fibrations.

We will use the following notation throughout. G is an
upper semicontinuous decomposition of S³ in which every
element has the shape of a circle, and f : S³ → S³/G is the

decomposition map. If $U \subset S^3/G$, then $\tilde{U}$ denotes $f^{-1}(U)$. Similarly if $x \in S^3/G$, then g_x denotes the decomposition element $f^{-1}(x)$ in S^3. Fundamental groups are denoted by π_1, and H_i or H^i denote integral, singular homology or cohomology groups. Also $\check{\pi}_1$, $\check{H}_i$, $\check{H}^i$ denote the corresponding Čech, alias shape, or inverse (direct) limit, groups.

2. Examples

Given any pair of relatively prime, non-negative integers p and q, let G_1 be the decomposition of $S^1 \times S^1$ whose elements are parallel (p,q) curves. (We allow $p = 0$ and $q = 1$, or $p = 1$ and $q = 0$.) Now consider S^3 to be the join of two circles Σ and Γ. There is a homeomorphism

$$h : S^1 \times S^1 \times (0,1) \to S^3 - (\Sigma \cup \Gamma) \quad .$$

Define a decomposition G_2 of S^3 by

$$G_2 = \{\Sigma, \Gamma\} \cup \{h(g \times \{t\}) \mid g \in G_1 , \quad 0 < t < 1\} \quad .$$

If $p = q = 1$, then the decomposition map $f_2 : S^3 \to S^3/G_2$ is a Hopf fibration [Ho]. If $p \neq 0$ and $q \neq 0$, then G_2 is a Seifert fiber space structure for S^3 [S]. Hypothesis i) of Theorem 1 is intended to prevent having to consider the case $p = 0$ or $q = 0$. Now let $k : S^3 \to S^3$ be any cellular mapping and let G be the decomposition of S^3 into the point inverses of $f_2 \circ k$. The elements of G have the shape of a circle and if $p \neq 0$, and $q \neq 0$ satisfy the other hypotheses of Theorem 1 as well. Of course S^3/G is homeomorphic to S^2 here by construction.

3. The Proof of Theorem 1

Most of the proof follows the proof of Theorem 1 of [CD$_3$]. We will outline this part of the argument in order to be able to

explain the concluding step. However for details we refer the
reader to $[CD_3]$ and $[CD_5]$.

Step 1. Winding functions. Fix $g \in G$. Since g has the
shape of S^1, g is an FANR $[B_2]$. Therefore [M] there are
saturated open neighborhoods $\widetilde{U}$ and $\widetilde{V}$ of g such that
$\pi_1 g \to \pi_1 \widetilde{U}$ is monic, and there is a homomorphism $r : \pi_1 \widetilde{V} \to \check{\pi}_1 g$
such that

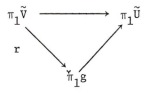

commutes where the unlabeled arrows are inclusion maps.
(Throughout we will mean inclusion maps by unlabeled arrows.)
Let $V = f(\widetilde{V})$ and define a map $\alpha_g : V \to R$ by $\alpha_g(x)$ is the
absolute value of the degree of the composition
$\check{\pi}_1 g_x \to \pi_1 \widetilde{V} \to \check{\pi}_1 g$. Note that by hypothesis i) we may assume
$\alpha_g(x) > 0$ for all x. When unambiguous we will write α
for α_g.

Step 2. The map α is lower semicontinuous. This follows
because $\alpha_g(x) = \alpha_g(y) \cdot \alpha_{g_y}(x) \geq \alpha_g(y)$ for any y near x.

Step 3. If $C = \{x \in V \mid \alpha$ is continuous at $x\}$, then C
is open and dense in V. If $D = V - C$, then $D = D_1 \cup D_2$
where D_2 is countable and D_1 is dense in itself (i.e.
every neighborhood of any point in D_1 contains uncountably
many points of D_1) [Bo, p. 395], [Ha, Ch. IX], or [MZ,
p. 262].

Step 4. $f \mid \tilde{C} : \tilde{C} \rightarrow C$ is an approximate fibration. This means that given any map $h : X \rightarrow \tilde{C}$ and any homotopy $H : X \times I \rightarrow C$ such that $fh = H \mid X \times \{0\}$ and any open cover ε of C, there is a homotopy $\tilde{H} : X \times I \rightarrow \tilde{C}$ such that $\tilde{H}(x,0) = h(x)$ and both $f\tilde{H}(x,t)$ and $H(x,t)$ lie in one element of ε for each $x \in X$, $t \in I$, $[CD_1]$. The key to this step is the fact that $\check{\pi}_1 g_x \rightarrow \check{\pi}_1 \tilde{V} \rightarrow \check{\pi}_1 g$ is an isomorphism for each $x \in C$. In other words $f \mid \tilde{C}$ is 1-movable. The result then follows from $[CD_2]$.

Step 5. Suppose A is an arc in V with $\text{Bd } A = \{c,d\}$ where $A - \{d\} \subset C$ and $\alpha(c) = p \cdot \alpha(d)$, $p \geq 1$. Then the inclusion $g_d \subset \tilde{A}$ is a shape equivalence and $\check{H}^1 \tilde{A} \rightarrow \check{H}^1 g_c$ is multiplication by p. The idea of the proof is that since A contracts to d holding d fixed, we can use the approximate fibration property to prove that $\tilde{A}$ shape deformation retracts to g_d. The second part of the conclusion follows from the definition of α, the Hurewicz Theorem and the Universal Coefficient Theorem.

Step 6. D is finite. If D_1 were not empty, we could find an arc A such that $\text{Bd } A = \{d_1, d_2\} \subset D$, $\text{Int } A \subset C$ and $\alpha(c) = p \cdot \alpha(d_1) = p \cdot \alpha(d_2)$ for $c \in \text{Int } A$ and $p > 1$. Using a Mayer-Vietoris sequence and duality we calculate $\tilde{H}_0(S^3 - A) \cong Z_p$. This is impossible, so $D = D_2$ is countable. Now C is connected so $\alpha(V)$ is bounded by the constant $\alpha(C)$. By the above argument the elements of D have different images under α; hence D is finite. In addition if $\alpha(C) = p_1 \cdot \alpha(d_1) = p_2 \cdot \alpha(d_2)$ where p_1 and p_2 have a least common divisor of $p > 1$, it still follows that $\tilde{H}_0(S^3 - A) \cong Z_p$. Hence the values $\alpha(d_i)$, $d_i \in D$, are relatively prime.

Step 7. S^3/G is homeomorphic to S^2. We use the Kline
Sphere Characterization [Bi]. It is well-known that S^3/G is
non-degenerate, locally connected, connected and metric. No
pair of points can separate S^3/G since by duality no two
disjoint sets with the shape of S^1 can separate S^3. It remains to be proven only that any simple closed curve separates
S^3/G.

Let J be a simple closed curve in S^3/G. Notice that d
is a discontinuity of α_g if and only if it is a discontinuity of α_g , whenever d is in the domain of both. Hence
there are only finitely many discontinuity points $d_1, \ldots, d_n$
on J. Also, the argument of Step 6 can be stretched across
the domains of winding functions, so the winding numbers
associated with the d_i's are still relatively prime. Choose
points of continuity $c_i \in J$ between d_i and d_{i+1}. (For
brevity a subscript of n+1 means a subscript of 1 throughout this argument.) Let A_{i+1} be the subarc between c_i and
c_{i+1} containing d_i; and let $g_i = f^{-1}(c_i)$ and
$h_i = f^{-1}(d_i)$. By the argument of Step 5, $h_i \subset \tilde{A}_i$ is a shape
equivalence. Likewise, $\check{H}^1\tilde{A}_i \to \check{H}^1 g_i$ and $\check{H}^1\tilde{A}_{i+1} \to \check{H}^1 g_i$ are
multiplication by p_i and p_{i+1} respectively where

$$\alpha_{h_i}(c_i) = p_i \quad \text{and} \quad \alpha_{h_{i+1}}(c_i) = p_{i+1} .$$

If i < n, we prove inductively that $H^1(\tilde{A}_1 \cup \cdots \cup \tilde{A}_i) \cong Z$,
$\check{H}^2(\tilde{A}_1 \cup \cdots \cup \tilde{A}_i) \cong 0$, and $H^1(\tilde{A}_1 \cup \cdots \cup \tilde{A}_i) \to \check{H}^1 g_i$ is
multiplication by $p_1 p_2 \cdots p_i$. Consider the Mayer-Vietories
sequence where $B_i = \tilde{A}_1 \cup \cdots \cup \tilde{A}_i$:

$$0 \leftarrow \check{H}^2\tilde{B}_i \leftarrow \check{H}^1 g_{i-1} \overset{\phi}{\leftarrow} \check{H}^1\tilde{B}_{i-1} \oplus \check{H}^1\tilde{A}_i \overset{\psi}{\leftarrow} \check{H}^1\tilde{B}_i \leftarrow 0 .$$

We have $\check{H}^1 g_{i-1} \cong \check{H}^1\tilde{B}_{i-1} \cong \check{H}^1\tilde{A}_1 \cong Z$ and $\phi(x,t) =$
$p_1 \cdots p_{i-1} s - p_i t$. Since $p_1 \cdots p_{i-1}$ is relatively prime to

p_i, ϕ is onto and $\check{H}^2\tilde{B}_i \cong 0$. Also $H^1B_i = \operatorname{im} \psi = \ker \phi = \{(s,t) \mid p_1 \cdots p_{i-1} s = p_i t\} \cong Z$ and $\psi(r) = (p_i r, p_1 \cdots p_{i-1} r)$. Hence $\check{H}^1\tilde{B}_i \to \check{H}^1A_i \to \check{H}^1g_i$ is multiplication by $p_1 p_2 \cdots p_i$.

Now consider the effect of adding the last arc A_n, thus closing J. Again the Mayer-Vietoris sequence yields

$$0 \leftarrow \check{H}^2\tilde{J} \leftarrow \check{H}^1g_{n-1} \oplus \check{H}^1g_n \overset{\phi}{\leftarrow} \check{H}^1\tilde{B}_{n-i} \oplus \check{H}^1A_n \leftarrow \cdots \quad .$$

Here $\check{H}^1g_{n-1} \cong \check{H}^1g_n \cong \check{H}^1\tilde{B}_{n-1} \cong \check{H}^1\tilde{A}_n \cong Z$ and $\phi(s,t) = (p_1 \cdots p_{n-1}s - p_n t, p_{n-1} p_1 s - p_n t)$. Hence $\check{H}^2\tilde{J} \cong Z$. Then by duality $\tilde{H}_0(S^3 - \tilde{J}) \cong Z$ so $\tilde{J}$ separates S^3. Thus J separates S^3/G since f is monotone. We therefore conclude that $S^3/G \cong S^2$. The second part of the conclusion, that f is approximable by Seifert fiber maps, follows from $[CD_3]$.

4. The Proof of Theorem 2 and Some Questions

By $[Cd_6]$ S^3/G is locally n-connected for all n. Hence by $[B_1$, Ch. V, Th. 103] S^3/G is an ANR. Therefore $S^3/G \cong S^2$ by Theorem 1.

Of course the main question remaining is:

Question 1. Let G be an upper semicontinuous decomposition of S^3 in which every element has the shape of a circle. Is $S^3/G \cong S^2$?

Short of answering this, one could try to remove either con-condition i) or ii) from Theorem 1 or 2. In particular removing condition ii) of Theorem 2 might be possible using the techniques of [KW]. In particular we ask:

Question 2. Let G be an upper semicontinuous decomposition of S^3 in which every element has the shape of a circle. Is S^3/G finite dimensional?

The first six steps of the proof of Theorem 1 work in other settings: for example with other 3-manifolds or with higher dimensional analogues, see $[CD_4]$. The seventh step doesn't generalize immediately, so we are led to ask:

Question 3. Let G be an upper semicontinuous decomposition of a 3-manifold M^3 in which every element has the shape of a circle. Is M^3/G a 2-manifold? Is the projection approximable by Seifert fiber maps? Can such decompositions exist for 3-manifolds which are not Seifert fiber spaces?

REFERENCES

[Bi]   Bing, R. H., "The Kline sphere characterization problem." Bull. Amer. Math. Soc. 52 (1946), 644-653.

$[B_1]$ Borsuk, K., Theory of Retracts. Polish Scientific Publishers, Warsaw, 1967.

$[B_2]$ ------, "A note on the theory of shape of compacta." Fund. Math. 67 (1970), 265-278.

[Bo]   Bourbaki, N., Elements of Mathematics: General Topology. Addison-Wesley, Reading, Mass., 1966.

$[CD_1]$ Coram, D. S. and Duvall, P. F., "Approximate fibrations." Rocky Mt. J. of Math. 7 (1977), 275-288.

$[CD_2]$ ------, "Approximate fibrations and a movability conditions for maps." Pac. J. Math. 72 (1977), 41-56.

$[CD_3]$ ------, "Mappings from S^3 to S^2 whose point inverses have the shape of a circle." Gen. Top. and Appl. 10 (1979), 239-246.

$[CD_4]$ ------, "Non-degenerate k-sphere mappings." Topology Proceedings 4 (1979), 67-82.

$[CD_5]$ ------, "Finiteness theorems for approximate fibrations." To appear.

$[CD_6]$ ------, "Local n-connectivity and approximate lifting." To appear.

[H]   Hausdorff, F., Set Theory. Chelsea, New York, 1957.

[Ho]   Hopf, H., "Uber die Abbildungen von Spharen auf Spharen
       niedriger Dimension." Fund. Math. 25 (1935), 427-440.
[KW]   Kozlowski, G. and Walsh, J., "The finite dimensionality
       of cell-like images of 3-manifolds." To appear.
[M]   Mardesic, S., "Strongly movable compacta and shape
       retracts." Proc. Internat. Symp. Top. Appl., Budva
       1972, 163-166.
[MZ]   Montgomery, D. and Zippin, L., Topological Transforma-
       tion Groups. Interscience, New York, 1955.
[S]   Seifert, H., "Topologie driedimensionaler gefaserter
       Raume." Acta Math. 60 (1963), 147-238.

A Mismatch Property in Spherical Decomposition Spaces

Robert J. Daverman

1. INTRODUCTION

Since 1977 almost all research concerning cell-like decompositions of n-manifolds (without boundary), when $n \geq 5$, has ex-
ploited R. D. Edwards' landmark Cell-Like Approximation
Theorem $[Ed_1]$, $[Ed_2, p.118]$, stated in one form as follows:

APPROXIMATION THEOREM: Let $f : M^n \to X$ denote a proper,
cell-like map of an n-manifold M^n $(n \geq 5)$ onto a finite
dimensional metric space X. Then f can be approximated by
homeomorphisms if and only if X has the Disjoint Disks
Property (any two maps of B^2 to X can be approximated by
ones having disjoint images).

By several standards this result has made studying decompositions of S^n, $n \geq 5$, simpler than those of S^3, despite
the potential for direct physical visualization and manipula-
tion with the latter.

To apply the Approximation Theorem to a particular cell-
like usc (upper semicontinuous) decomposition G of an
n-manifold M, two problems must be addressed; the first and
apparently harder is to detect the finite dimensionality of
M/G, and the second is to determine whether M/G satisfies
the Disjoint Disks Property (henceforth to be abbreviated as
DDP). In this paper we reconsider a property of decomposition
spaces that immediately dispels the first problem. When G
is a cell-like usc decomposition of an n-manifold M, we say
that M/G is *spherical* if each $x \in M/G$ has arbitrarily
small neighborhoods U_x whose frontiers $Fr\ U_x$ are

(n - 1)-spheres. Under such circumstances, if M/G is
spherical, then clearly it is a finite dimensional space;
whether it satisfies the DDP when $n \geq 5$ remains unsolved.
However, evidence from lower dimensional manifolds suggests
that it might not, for S. Armentrout [A] has shown that the
decomposition space associated with his decomposition of E^3
into points and straight line segments is spherical even though
it is not a manifold.

There are two earlier results pertaining to spherical cell-
like decompositions G of n-manifolds M $(n \geq 5)$. One is due
to J.W. Cannon [C, Theorem 62], who showed that $\pi: M \to M/G$
can be approximated by homeomorphisms if the (n - 1)-spheres
Fr U_x promised by the sphericality hypothesis are 1-LCC embedded in M/G. The other is by the author [D, Theorem 4.4],
who obtained the same conclusion if the neighborhoods U_x are
1-LC at the points of Fr U_x. (See the papers cited for
definitions.)

Here we trade the strict 1-LC property on sides of the
(n - 1)-sphere Fr U for a more relaxed mismatch property.
An (n - 1)-sphere Σ is the decomposition space M/G (associated as usual with a cell-like usc decomposition of a connected n-manifold M) is said to satisfy the *Homotopy Mismatch
Property* (abbreviated HMP) if Σ contains disjoint subsets
Q_1 and Q_2 such that any map $f_i : B^2 \to \text{Cl } U_i$ can be approximated by a map $f'_i : B^2 \to U_i \cup Q_i$ (i = 1, 2), where U_1 and
U_2 denote the components of $(M/G) - \Sigma$. The main result of
this paper is the following.

THEOREM 1. Suppose G is a cell-like usc decomposition
of an n-manifold M, $n \geq 5$, such that each $x \in M/G$ has arbitrarily small neighborhoods U_x whose frontiers Fr U_x satisfy the HMP. Then $\pi: M \to M/G$ can be approximated by homeomorphisms.

(As is customary, N_G denotes the union of the nondegenerate sets g from G.) Theorem 1 answers a question of W.T. Eaton (cf. [C, p. 101]). A corollary subsumes both earlier theorems about spherical decompositions in [C] and [D].

COROLLARY A. Suppose G is a cell-like usc decomposition of an n-manifold M, n $\geq$ 5, such that each x $\in$ M/G has arbitrarily small neighborhoods U_x for which either U_x or (M/G) - Cl U_x is 1-LC at points of Fr U_x. Then π : M $\to$ M/G can be approximated by homeomorphisms.

Work on these results was inspired by a recent paper of E. Woodruff [W_2] about decompositions of S^3. This one extends hers to higher dimensions and generalizes somewhat, for her hypotheses describe geometry, in the source, of the decomposition itself rather than of the decomposition space. Nevertheless, her hypotheses do generate neighborhoods U_x in the decomposition space whose frontiers do satisfy the HMP, as in Theorem 1. This work was also spurred through several conversations with J.P. Henderson, to whom the author wishes to express his appreciation.

Theorem 1 also establishes a high dimensional analogue of another Woodruff result [W_1].

COROLLARY B. Suppose Ġ is a cell-like usc decomposition of an n-manifold M, n $\geq$ 5, such that each nondegenerate element g of G has arbitrarily small neighborhoods V_g in M whose frontiers Fr V_g miss N_G and satisfy the HMP in M. Then π: M $\to$ M/G can be approximated by homeomorphisms.

In the 3-dimensional situation investigated by Woodruff, Eaton's Mismatch Theorem [Ea_1, Ea_2] implies that the 2-sphere

frontiers, as above, must satisfy the HMP, but in higher dimensions (n - 1)-spheres do not necessarily satisfy it. Whether
Corollary B is valid without the hypothesis that the frontiers
satisfy the HMP stands as an open question.

2. PROOF OF THEOREM 1.

   LEMMA 1. Suppose G is a cell-like usc decomposition of
an n-manifold M, $n \geq 5$, such that M/G is spherical and the
$(n - 1)$-spheres Fr U satisfy the HMP; f_1 and f_2 maps of
B^2 to M/G; C a closed subset of M/G such that
$C_0 = C \cap f_1(B^2) \cap f_2(B^2)$ is 0-dimensional; and W an open
subset of M/G containing C_0. Then there exist maps
F_1, $F_2 : B^2 \to M/G$ satisfying

   (1) $F_1(B^2) \cap F_2(B^2) \cap C = \emptyset$,

   (2) $F_1 f_i^{-1}(W) \subset W$,   and

   (3) $F_i | B^2 - f_i^{-1}(W) = f_i | B^2 - f_i^{-1}(W)$ $(i = 1, 2)$.

   Proof. We break the proof into several steps.

   Step 1: *Finiteness Considerations.* By hypothesis each
point $c \in C_0$ has an open neighborhood U_c whose closure is
in W and whose frontier Fr U_c is an $(n - 1)$-sphere satisfying the HMP. Since C_0 is compact we can extract from the
open cover $\{U_c | c \in C_0\}$ a finite subcover $\{U_j | 1 \leq j \leq r\}$
and trim the latter to another cover $\{V_j | 1 \leq j \leq r\}$ of C_0
by open sets in M/G such that $V_j \subset Cl\ V_j \subset U_j$.

   Step 2: *A Reduction.* Given maps f_1, $f_2 : B^2 \to M/G$ for
which

$$f_1(B^2) \cap f_2(B^2) \cap C \subset \bigcup_{j=k}^{r} V_j \ ,$$

we shall describe maps F_1, $F_2 : B^2 \to M/G$ such that

(1') $F_1(B^2) \cap F_2(B^2) \cap C \subset \bigcup_{j=k+1}^{r} V_j$,

(2') $F_i \, f_i^{-1}(W) \subset W$; and

(3') $F_i | B^2 - f_i^{-1}(W) = f_i | B^2 - f_i^{-1}(W)$ $(i = 1, 2)$.

Repeated application (r times) of this reduced version will
establish Lemma 1.

Step 3: *Eradication of* $f_i(B^2)$ *from* U_k. Define
$Z_i = f_i^{-1} (Cl \ U_k)$ and $Y_i = f_i^{-1}(Fr \ U_k)$ $(i = 1, 2)$. The map
$f_i | Y_i : Y_i \to Fr \ U_k$ extends to a map $m_i : Z_i \to Fr \ U_k$ $(i = 1,2)$,
since $Fr \ U_k$ is a simply connected ANR.

Step 4: *General Position Improvements*. To circumvent
special difficulties when $n = 5$, we consider only the case
$n \geq 6$, leaving the extra case to the reader. Because
$\dim Fr \ U_k \geq 5$, m_1 and m_2 admit general position modifications, affecting no points of Y_1 or Y_2, so that
$m_1(Z_1 - Y_1) \cap m_2(Z_2 - Y_2) = \emptyset$. Define maps $f_i' : B^2 \to M/G$ as
m_i on Z_i and as f_i elsewhere $(i = 1, 2)$. Then any
$c \in f_1'(B^2) \cap f_2'(B^2) \cap C$ belongs either to

$$f_1(B^2) \cap f_2(B^2) - V_k \subset f_1(B^2) \cap f_2(B^2) \cap (\bigcup_{j=k+1}^{r} V_j)$$

or to

$$C \cap ([f_1(Y_1) \cap m_2(Z_1 - Y_2)] \cup [m_1(Z_1 - Y_1) \cap f_2(Y_2)]) .$$

Only the points of the second kind cause any further concern.

Step 5: *Final improvements to* m_i. Define sets

$$X_1 = Z_1 \cap m_1^{-1}(C \cap m_1(Z_1) \cap f_2(Y_2)) \quad \text{and}$$
$$X_1^* = X_1 - m_1^{-1}(\bigcup_{j=k+1}^{r} V_j)$$

and define sets X_2 and X_2^* symmetrically. Then X_i^* is a compact subset of X_i and $X_i^* \subset Z_i - Y_i$ $(i = 1, 2)$, for if $x \in X_1^* \cap Y_1$, then

$$m_1(x) \in [C \cap m_1(Y_1) \cap f_2(Y_2)]$$

$$- \bigcup_{j=k}^{r} V_j \subset [C \cap f_1(Y_1) \cap f_2(Y_2)] - \bigcup_{j=k}^{r} V_j$$

which is empty by the reductive hypothesis of Step 2. Define $T_1 = f_1^{-1} m_2(X_2^*)$ and $T_2 = f_2^{-1} m_1(X_1^*)$. It should be clear that $T_i \subset Y_i$ $(i = 1, 2)$ and that

$$m_1(X_1^*) \cap m_2(X_2^*) = \emptyset = f_1(T_1) \cap f_2(T_2) .$$

Next determine open sets N_i and O_i in B^2 such that

$$T_i \subset N_i \subset f_i^{-1}(W) , \ X_i \subset O_i \subset Cl \ O_i \subset Z_i - Y_i , \ N_i \cap O_i = \emptyset$$

$$(i = 1, 2)$$

and

$$f_1'(Cl \ N_1) \cap f_2'(Cl \ N_2) = \emptyset = f_1'(Cl \ O_1) \cap f_2'(Cl \ O_2) .$$

Since $f_1'(B^2) \cap f_2'(B^2) \subset f_1'(X_1^*) \cup f_2'(X_2^*) \cup (\bigcup_{j=k+1}^{r} V_j)$, it follows that

$$[f_1'(B^2) \cap f_2'(B^2 - O_2)] \cup [f_1'(B^2 - O_1) \cap f_2'(B^2)] \subset \bigcup_{j=k+1}^{r} V_j .$$

Let Q_1 and Q_2 denote disjoint sets in $Fr \ U_k$, promised by the hypothesis that $Fr \ U_k$ satisfies the HMP, such that $U_k \cup Q_1$ and $(M/G - Cl \ U_k) \cup Q_2$ are 1-LC at points of $Fr \ U_k$. Then one can produce maps $F_i' : Cl(N_i \cup O_i) \to M/G$ such that

$$F_i'(Cl \ N_i) \subset (W - Cl \ U_k) \cup Q_2 ;$$

$$F_i'(Cl \ O_i) \subset U_k \cup Q_1 , \quad \text{and}$$

$$F_i'|Fr \ N_i \cup Fr \ O_i = f_i'|Fr \ N_i \cup Fr \ O_i \quad (i = 1, 2)$$

with controls on the closeness of F'_i to f'_i $(i = 1, 2)$ so that

$$F'_1(\text{Cl } 0_1) \cap F'_2(\text{Cl } 0_2) = \emptyset \quad \text{and}$$

$$[F'_1(\text{Cl } N_1) \cap f'_2(B^2 - 0_2)] \cup [f'_1(B^2 - 0_1) \cap F'_2(\text{Cl } N_2)] \subset \bigcup_{j=k+1}^{r} V_j .$$

Consequently, the maps F_i defined as F'_i on $N_i \cup 0_i$ and as f'_i elsewhere fulfill the requirements for the reduction given in Step 2.

LEMMA 2. Suppose G is a cell-like usc decomposition of an n-manifold M, $n \geq 5$, such that M/G is spherical and the $(n - 1)$-spheres Fr U satisfy the HMP; f_1 and f_2 maps of B^2 to M/G; $C \subset M/G$ a closed q-dimensional subset of M/G; W an open subset of M/G containing $f_1(B^2) \cap f_2(B^2) \cap C$; and $\varepsilon > 0$. Then there exist maps F_1, $F_2 : B^2 \to M/G$ satisfying

(1) $F_1(B^2) \cap F_2(B^2) \cap C = \emptyset$,

(2) $\rho(F_i, f_i) < \varepsilon$, and

(3) $F_i | B^2 - f_i^{-1}(W) = f_i | B^2 - f_i^{-1}(W)$ $(i = 1, 2)$.

Proof. The argument proceeds by induction on q, and Lemma 1 establishes the initial case $q = 0$. Assume that Lemma 2 holds for all $(q - 1)$-dimensional closed subsets of M/G. In the q-dimensional set C find closed subsets A_1, $A_2, \cdots$ of dimension $\leq q - 1$ such that $C - \cup A_j$ is 0-dimensional. Apply the inductive hypothesis repeatedly, imposing controls with an eye toward the limit, to obtain (limit) maps $f_i^* : B^2 \to M/G$ such that $f_i^*(B^2) \cap (\cup A_j) = \emptyset$, $\rho(f_i^*, f_i) < \varepsilon/2$, and

$$f_i^* | B^2 - f_i^{-1}(W) = f_i | B^2 - f_i^{-1}(W) \quad (i = 1, 2) .$$

Then $C \cap f_1^*(B^2) \cap f_2^*(B^2)$ is 0-dimensional, since it is a
subset of $C - \cup A_j$.   Now apply Lemma 1 again, this time with
open set W^* in W containing $C \cap f_1^*(B^2) \cap f_2^*(B^2)$ and
having no component of diameter $\geq \varepsilon/2$,   to obtain the required maps F_1 and F_2.

   When $q = n$ and $C = M/G$,   Lemma 2 reveals that   M/G
satisfies the DDP.   Theorem 1 follows from Edwards' Approximation Theorem $[Ed_2]$.

   Stripping the argument to its essentials, one discovers
that it proves the following result:

   THEOREM 2.   Suppose   G   is a cell-like   usc   decomposition
of an n-manifold, $n \geq 5$,   such that each point   x   of $\pi(N_G)$
has arbitrarily small neighborhoods U_x such that   Fr U_x is
a simply connected   (n - 1)-manifold or a simply connected
ANR with the DDP and   Fr U_x satisfies the HMP.   Then
$\pi : M \to M/G$ can be approximated by homeomorphisms.

ACKNOWLEDGMENT

This research was supported in part by the National Science
Foundation under Grant MCS 79-06083.

REFERENCES

[A]   S. Armentrout, "A three-dimensional spheroidal space
       which is not a sphere," Fund. Math. 68 (1970), 183-186.

[C]   J. W. Cannon, "Taming cell-like embedding relations," in
       Geometric Topology (L. C. Glaser and T. B. Rushing, eds.)
       Lect. Notes in Math. #453, Springer-Verlag, New York,
       1975, pp. 66-118.

[D]   R. J. Daverman, "Detecting the disjoint disks property,"
       Pacific J. Math., to appear.

[Ea$_1$] W. T. Eaton, "Sums of solid spheres," Michigan Math. J.
       19 (1972), 193-207.

[Ea$_2$] ------, "A 2-sided approximation theorem for 2-spheres,"
       Pacific J. Math. 44 (1973), 461-485.

[Ed$_1$] R. D. Edwards, "Approximating certain cell-like maps by
       homeomorphisms," manuscript.

[Ed$_2$] ------, "The topology of manifolds and cell-like maps,"
       in Proceedings of the International Congress of Mathe-
       maticians, Helsinki, 1978, Academia Scientarium Fennica,
       Helsinki, 1980, pp. 111-127.

[W$_1$] E. P. Woodruff, "Decomposition spaces having arbitrarily
       small neighborhoods with 2-sphere boundaries," Trans.
       Amer. Math. Soc. 232 (1977), 195-204.

[W$_2$] ------, "Decomposition spaces having arbitrarily small
       neighborhoods with 2-sphere boundaries II," preprint.

Countable Starlike Decompositions of S³

Richard Denman

The purpose of this article is to provide another proof of the following result of Bing [B]. Recently, this result has been extended to include countable starlike equivalent decompositions and more.

THEOREM: If G is an upper semicontinuous decomposition of S^3 with countably many non-degenerate elements, each of which is starlike, then S^3/G is homeomorphic to S^3.

Let $N(G) = \{g(n): n \in \omega\}$ denote the set of non-degenerate elements of G. By the Bing shrinking criterion, it will suffice to show that given a saturated open set U containing $N(G)$ and given $\varepsilon > 0$, there is a homeomorphism $H: S^3 \to S^3$ such that $H/(S^3 - U) =$ identity and diam $(H(g)) < \varepsilon$ for each $g \in N(G)$. The main tool that will be used is Lemma 1 which shows, as in the original proof, how to shrink one non-degenerate element toward its star point, while limiting the growth of all others. For a proof, see [1].

LEMMA 1: Let G be as above, let $\varepsilon > 0$, and let $g(n) \in N(G)$. Also, let $\{C(1,n), C(2,n), \cdots, C(K(n),n)\}$ be a nested collection of ideally starlike 3-cells about $g(n)$ such that $K(n) \cdot \varepsilon/4 > rad(g(n))$. Assume, by upper semicontinuity, that no non-degenerate element intersects the boundary of more than one of the $C(i,n)$. Then there is a homeomorphism $h(n): S^3 \to S^3$ such that:

   a) $diam(h(n)(C(i,n))) < (K(n) - i + 1)\varepsilon/2$.

b) $\text{diam}(h(n)(g)) < \text{diam}(g) + \varepsilon/2$ for each $g \in N(G)$.

c) $h | (S^3 - \text{Int } C(1,n)) = \text{id}$.

If Lemma 1 is applied repeatedly without care in choosing the
defining cells, some non-degenerate element may grow quite
large, even though its growth is controlled in each single
application. Therefore, before any shrinking is done, the
defining cells are chosen carefully to avoid this growth.
Lemma 2 describes how to choose $\{C(1,n), \cdots, C(K(n),n)\}$ about
$g(n)$ after the cells have been chosen for all the previous
$g(i)$. The proof is a straightforward application of upper
semicontinuity.

LEMMA 2: Let U, G, and ε be as above. Then there is
a nested collection $\{C(1,n), \cdots, C(K(n),n)\}$ of ideally starlike 3-cells about $g(n)$ such that:

a) $C(1,n) \subset U$.

b) $K(n) \cdot \varepsilon/2^{n+3} > \text{rad}(g(n))$.

c) If $m < n$, then $g(m) \cap C(1,n) = \phi$ and $C(1,n)$
intersects the boundary of at most one $C(j,m)$.

d) If $m < n$, $g \in N(G)$, $g(n) \cap C(j,m) = \phi$, and
$g \cap C(j,m) \neq \phi$, then $g \cap C(1,n) = \phi$.

Proof of Theorem. Notice that $\{C(K(n),n): n \in w\}$ is an
open cover of the compact set $B = \{g \in N(G): \text{diam}(g) \geq \varepsilon/2\}$,
so that for some $j \in \omega$, $B \subset \bigcup_{i=1}^{j} C(K(i),i)$. Now, define
$h(i)$ to be the homeomorphism of Lemma 1 corresponding to
$g(i)$, $\varepsilon/2^i$, and $\{C(1,i), \cdots, C(K(i),i)\}$ and define
$H = h(1) \circ h(2) \circ \cdots \circ h(j)$. Since the order in which these
first j non-degenerate elements are shrunk is the reverse
of the order in which their cells were chosen, conclusion (c)
of Lemma 2 guarantees that at each stage $h(i)$ of the

shrinking the non-degenerate elements with smaller subscripts remain starlike. Therefore, these elements can be used to guide the later stages of the shrinking. H is clearly the identity on $S^3 - U$, so it remains to show that $\text{diam}(H(g)) < \varepsilon$ for all $g \in N(G)$. Consider the following two cases:

Case 1. If $\text{diam}(g) < \varepsilon/2$, then by Lemma 1, the diameter of g grows less than $\varepsilon/2^{i+1}$ under $h(i)$, so that its total growth in diameter is less than $\sum_{i=1}^{j} \varepsilon/2^{i+1}$, which is less than $\varepsilon/2$.

Case 2. If $\text{diam } g \geq \varepsilon/2$, then $g \subset C(K(i),i)$ for some $i < j$. Now, at some stage $h(\ell)$ of the shrinking prior to $h(i)$ (that is, $i < \ell \leq j$) it is possible that $h(\ell)(g)$ is no longer inside $C(K(i),i)$. The figure illustrates how this might happen.

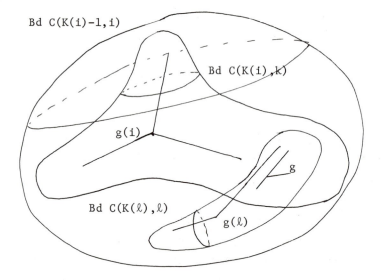

However, by conclusion (d) of Lemma 2, the image of g does
remain in C(K(i) -1,i) at each stage of the shrinking prior
to h(i). Now, by conclusion (a) of Lemma 1, h(i) shrinks
the diameter of C(K(i) -1,i) and hence the diameter of
$h_{i+1} \circ \cdots \circ h_j (g)$ to less than $\varepsilon / 2^i$, which is less than
$\varepsilon/2$. This puts $h_i \circ h_{i+1} \circ \cdots \circ h_j (g)$ in the category of
case 1, and the proof is complete.

REFERENCES

[B] R. H. Bing, "Upper semicontinuous decompositions of E^3."
    Ann. of Math. 65 (1957), 363-374.

General Position Properties Related to the Disjoint Discs Property

Dennis J. Garity

<u>0</u>. Recent work of J. W. Cannon [Ca2] and R. D. Edwards [Ed] has focused a great deal of attention on a specific general position property and its relationship to decompositions of manifolds of dimension greater than or equal to 5. This property is the Disjoint Discs Property, or DDP. The above mentioned work, along with that of F. Quinn [Qu], led to a surprising characterization of topological manifolds in this dimension range. Such manifolds are precisely those homology manifolds satisfying the minimal amount of general position provided by the DDP. Cannon details the history of these unexpected results in [Ca1].

In the following, we define general position properties closely related to the DDP and state some consequences of these properties. We also construct examples of decomposition spaces satisfying some but not other of these properties.

We consider cell-like (CE) upper semicontinuous (usc) decompositions of n-manifolds M, n $\geq$ 5. If G is such a decomposition, H_G denotes the set whose elements are the nondegenerate elements of G, N_G denotes the union of these elements, and π_G (or π if there is just one decomposition under consideration) denotes the natural quotient map from M onto M/G.

<u>1</u>. <u>DEFINITION</u>. A metric space X satisfies property DD_k, k $\geq$ 2, if given maps $f_1, \cdots, f_k$ from B^2 into X and $\varepsilon > 0$, there exist maps $g_1, \cdots, g_k$ from B^2 into X such that:

(i) $|f_i - g_i| < \varepsilon$ for $1 \le i \le k$;

and (ii) $\bigcap\limits_{i=1}^{k} g_i(B^2) = \phi$.

Note: Property DD_2 is the same as DDP.

DEFINITION. Let G be a CE usc decomposition of an n-manifold M, $n > 3$. Then G is said to be secretly d-dimensional if there exists a CE usc decomposition H of M, with $M/G \cong M/H$, and with the dimension of π_H (N_H) less than or equal to d.

Using the above terminology, the CE Approximation Theorem of Edwards can be stated as follows:

CE APPROXIMATION THEOREM. [Ed] Let G be a CE usc decomposition of an n-manifold M, $n \ge 5$, such that M/G is an ANR. Then M/G satisfies the DDP if and only if G is secretly (-1)-dimensional.

The statement of the following theorem parallels the statement of the above theorem. Its proof will be presented elsewhere.

THEOREM 1. Let G be a CE usc decomposition of an n-manifold M, $n \ge 5$, such that M/G is an ANR. If M/G satisfies DD_k, then G is secretly (k-3)-dimensional.

Note: The examples presented in the next section show that the converse to the above theorem is false.

2.1. It can be shown that decompositions satisfying any of the properties DD_k are cellular decompositions. If all cellular decompositions were secretly 0-dimensional, there would be little substance to Theorem 1. This is not the case.

See [D-G]. If all cellular decompositions yielding nonmani-
folds satisfied the simplest possible DD_k property, DD_3,
there would be little point in discussing the remaining properties. This section provides examples to overcome this possible objection. Specifically, the following theorem is shown
to be true.

THEOREM 2. There exist cellular usc decompositions G
of E^n, $n \geq 3$, satisfying DD_{k+1} but not DD_k, $k \geq 3$. If
$n \geq 4$, k can be taken to be equal to 2.

The examples constructed to prove Theorem 2 are closely
related to Bing's dogbone space [Bi] and to Eaton's generalization of that space [Ea]. A general technique developed by
Daverman and Edwards [Da, pg. 147] will be used. Where Eaton
uses 2 wild Cantor Sets, we will use k.

The following definition and lemma are taken from [Da].

DEFINITION. Let M be a manifold with boundary, H a
disc with holes, and g a map from H into M with
$g(\partial H) \subset \partial M$. Then g is said to be I-inessential if there
exists a map h from H into ∂M with $h|\partial H = g|\partial H$. Other-
wise, g is said to be I-essential.

LEMMA 1. [Da, pg. 147] Let S denote a closed P.L.
$(n-2)$-manifold and $N = S \times B^2$. Choose $\varepsilon > 0$. Then there
exists a finite collection $\{N_i\}$ of pairwise disjoint locally
flat manifolds in Int(N) such that:

   (i)   each N_i is homeomorphic to the product of B^2
         and a closed P.L. $(n-2)$-manifold;

   (ii) the diameter of each N_i is less than ε;

and (iii) whenever H is a disc with holes and g from H into M is an I-essential map, then

$$g(H) \cap (\cup N_i) \neq \phi \ .$$

In the construction to follow, it is assumed that the reader is familiar with the examples of Blankinship [Bk] and Eaton [Ea]. Implicit in their specific constructions is the following improvement to Lemma 1.

LEMMA 2. If the N from Lemma 1 is homeomorphic to the product of B^2 with (n-2) copies of S^1 , then the collection $\{N_i\}$ can be chosen as in Lemma 1 so that in addition:
    (iv) each N_i is homeomorphic to N;
and   (v) whenever g is a P.L. map from a disc with holes
          H into N with $g(\partial H) \subset \partial N$, and R is a spe-
          cific element from $\{N_i\}$, then there exists a
          map h from H into N with $h|\partial H = g|\partial H$, so
          that $g(H) \cap N_i = \phi$ unless $N_i = R$.

2.2. The Construction. The decomposition G of E^n will have as nondegenerate elements the components of $\cap M_i$ where each M_i is a finite collection of pairwise disjoint P.L. manifolds in E^n . Each nondegenerate element of G will be a k-od.

DEFINITION. Let $T^n \equiv B^2 \times S^1(1) \times \cdots \times S^1(n-2)$ be the product of B^2 with (n-2) copies of S^1 .

Stage 1
Let $e_1, \cdots, e_k$ be pairwise disjoint P.L. embeddings of T^n into E^n . Choose a point p in $E^n - \bigcup_{i=1}^{k} e_i(T^n)$. Let c be a P.L. k-od joining p to each $e_i(\partial T^n)$, with p as the wedge point. Let N be a regular neighborhood of c in E^n .

Define S to be $N \cup (\bigcup_{i=1}^{k} e_i(T^n))$ and let M_1 have S as its only component. Each component of each M_i will be homeomorphic to S. Figure 1 depicts the case $k = 4$ in E^3.

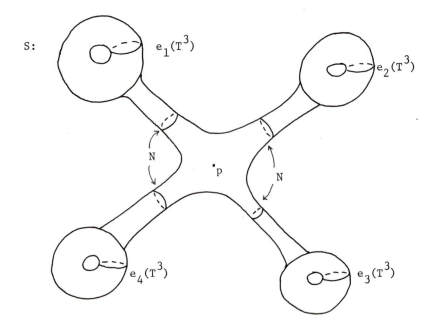

Figure 1

Consider the following two properties:

Property a_r. Let H_i, $1 \leq i \leq k$, be discs with holes and $g_i: H_i \rightarrow e_i(T^n)$ be P.L. I-essential maps in general position with respect to all components of M_p for $p \leq r$. Then for some component C of M_r, $C \cap g_i(H_i) \neq \phi$ for $1 \leq i \leq k$.

Property b_r. Let H_i, $1 \leq i \leq k+1$, be discs with holes and $g_i: H_i \to E^n$ be P.L. maps in general position with respect to all components of M_p for $p \leq r$. Then there exist maps $h_i: H_i \to E^n$, $1 \leq i \leq k+1$, with

$$h_i | \{g_i^{-1}(E^n - M_{r-1})\} = g_i | \{g_i^{-1}(E^n - M_{r-1})\} \; ;$$

with $h_i(g_i^{-1}(M_{r-1})) \subset M_{r-1}$; and with no component C of M_r having the property that

$$C \cap h_i(H_i) \neq \phi \qquad \text{for } 1 \leq i \leq k+1 \; .$$

For property b_r to make sense in the case $r = 1$, let M_0 be a large n-cell containing S. Then, M_1 can be easily seen to satisfy a_1 and b_1.

Stage j.

Assume M_{j-1} has been constructed so that each of its components is homeomorphic to S, and so that properties a_{j-1} and b_{j-1} hold. The next stage, M_j, will be constructed in the interior of M_{j-1} so that each component of M_j is homeomorphic to S and is contained in an n-cell in the interior of M_{j-1}. In addition, the construction will be done so that properties a_j and b_j hold.

Consider a specific component C of M_{j-1}. Let f be a homeomorphism from $S = N \cup (\bigcup_{i=1}^{k} e_i(T^n))$ onto C and $g_i = f \circ e_i$. Use Lemma 1 and Lemma 2 to find a finite collection $\{T(i,1), \cdots, T(i,q)\}$ of locally flat pairwise disjoint copies of T^n inside $g_i(T^n)$ for $1 \leq i \leq k$ so that:

   (i)   each $T(i,r)$ is of diameter less than $\varepsilon/2^j$ and is contained in an n-cell in the interior of $g_i(T^n)$;

(ii)   whenever H is a disc with holes and
       g: $H \to g_i(T^n)$ is an I-essential map, then
       $g(H) \cap (\bigcup_{r=1}^{q} T(i,r)) \neq \phi$;

and (iii)   if R is a specific element from $\{T(i,1), \cdots,$
       $T(i,q)\}$ and g is a P.L. map from a disc with
       holes H into $g_i(T^n)$ with $g(\partial H) \subseteq g_i(\partial T^n)$,
       then there exists a map h: $H \to g_i(T^n)$ with
       $h(H) \cap T(i,r) = \phi$ unless $T(i,r) = R$.

<u>Ramifying the copies of T^n.</u> If $T(i,r)$ is thought of as
$B^2 \times S^1(1) \times \cdots \times S^1(n-2)$, a parallel interior copy of
$T(i,r)$ is any subset of $T(i,r)$ of the form

$$\{\text{interior subdisc of } B^2\} \times S^1(1) \times \cdots \times S^1(n-2) .$$

The above properties allow one to choose pairwise disjoint
parallel interior copies of the $T(i,r)$ and to tube these
copies together in the interior of C in such a way so that
a collection of copies of S are formed. Stage M_j consists
of the union, over all the components of M_{j-1}, of these new
copies of S.

If sufficient care is taken, properties (ii) and (iii)
above allow one to do the above tubing operation so that properties a_j and b_j are satisfied.

The decomposition G

G is obtained by taking as nondegenerate elements the components of

$$\bigcap_{i=0}^{\infty} M_i .$$

It is routine to verify that G is a cellular usc decomposition of E^n.

2.3. Further discussion of Theorem 2. The fact that properties a_r and b_r hold for $r \in z^+$ allows one to show that
E^n/G satisfies DD_{k+1}, but not DD_k. The reason that k is
required to be greater than 2 when n is equal to 3 is
that three B^2 images cannot always be adjusted by general
position so as to have empty intersection in E^3.

By choosing a countable null sequence $\{A_1, A_2, \cdots\}$ of
n-cells in E^n and by using the construction of section 2.2
to form cellular decompositions of A_i that fail to satisfy
DD_{i+1}, one can prove the following corollary.

COROLLARY 1. There exist cellular usc decompositions G
of E^n, $n \geq 3$, that fail to satisfy property DD_k for all
finite k.

REFERENCES
[Bi]   Bing, R. H., "A decomposition of E^3 into points and
       tame arcs such that the decomposition space is topo-
       logically different from E^3," Ann. of Math. (2) 65
       (1957), 484–500.

[Bk]   Blankinship, W. A., "Generalization of a construction
       of Antoine," Ann. of Math. 53 (1951), 276–297.

[Ca1] Cannon, J. W., "The characterization of topological
       manifolds of dimension $n \geq 5$," Proceedings of the
       International Congress of Mathematicians, Helsinki,
       1978.

[Ca2] ------, "Shrinking cell-like decompositions of mani-
       folds. Codimension three," Ann. of Math. 110 (1979),
       83–112.

[Da]   Daverman, R. J., "On the absence of tame disks in cer-
       tain wild cells," in General Topology (L. C. Glaser and

T. B. Rushing, editors), Lecture notes in Math #438, Springer-Verlag, New York, 1975, 142-155.

[D-G] Daverman, R. J., and Garity, D. J., "Intrinsically (n - 2) dimensional cellular decompositions of n-manifolds, (n $\geq$ 3)." See Notices Amer. Math. Soc. 26 (1979), A-621 #79T-G110.

[Ea] Eaton, W. T., "A generalization of the dog bone space to E^n." Proc. Amer. Math. Soc. 39 (1973), 379-387.

[Ed] Edwards, R. D., "Approximating certain cell-like maps by homeomorphisms," preprint. See Notices Amer. Math. Soc. 24 (1977), A-649, #751-G5.

[Qu] Quinn, F., "Ends of maps, I." Ann. of Math. 11 (1979), 275-331.

A Discussion of Results and Problems Related to Cellularity in Polyhedra

James P. Henderson

1. INTRODUCTION

The concept of a cellular subset X of a manifold M^n has played an important role in geometric topology over the past 20 or more years. In 1960, Brown [4] defined a set X to be cellular in an n-manifold M^n if there are n-cells Q_1, $Q_2, \ldots$ in M^n such that $Q_{i+1} \subset \text{int } Q_i$ and $\overset{\infty}{\underset{i=1}{\cap}} Q_i = X$. In that paper, he used this concept in his proof of the generalized Schoenflies theorem. Later, McMillan [10] was able to give a criterion for determining which cell-like subsets in the interior of a P.L. n-manifold, $n \geq 5$, were in fact cellular subsets of that manifold.

Also studied in great detail were those proper surjections $f : M^n \to N^n$ between n-manifolds such that for each $y \in N$, $f^{-1}(y)$ is a cellular subset of M. Armentrout [2] showed that such maps are approximable by homeomorphisms if $n = 3$. Siebenmann [12] then proved that for $n \neq 4$ such maps are approximable by homeomorphisms.

In [6], Cannon proposed a generalization of the concept of cellularity to polyhedra. He defined a compact subset X of a polyhedron P to be *cellular* in P if there is a pseudo-isotopy $g_t : P \to P$ such that X is the only nondegenerate point preimage of g_1. It should be noted that if we replace polyhedron by manifold in the above definition we have a definition of cellularity in a manifold which is equivalent to that of Brown.

As an example, let P be the wedge of a 2-simplex σ and a 1-simplex τ at a vertex of each. Denote the wedge point

by v. If we consider τ as a subset of P, τ is not a
cellular subset. Similarly, σ is not cellular in P. However, if we let X be a closed subinterval of τ containing
exactly one of the endpoints of τ, then X is cellular in
P. Any embedded arc A whose interior lies in the interior
of σ and whose endpoints lie in the boundary of σ is not
cellular in P. If X is an embedded arc in σ such that
X intersects the boundary of σ in a single point, then X
is cellular in P.

We now turn to a survey of results and questions related
to cellularity in polyhedra and some of the techniques used
to deal with these problems. Throughout, P and Q will be
finite dimensional polyhedra.

2. STRATIFICATION

In order to be able to understand cellularity in polyhedra,
we must first break our polyhedra down into a collection of
pairwise disjoint manifolds of various dimensions. This idea
of stratification of a polyhedron has been used by Aiken [1],
Handel [7], and others [3, 13]. It should be pointed out that
Aiken required that the manifolds of his stratification inherit a P.L. manifold structure from P, which we will not
normally do.

The *intrinsic dimension* of a point x in P, denoted
I(x, P), is given by I(x, P) = max$\{n \in \mathbb{Z} \mid$ there is an open
embedding $h : \mathbb{R}^n \times cL \to P$, where cL is the open cone on a
compact polyhedron L and $h(0 \times c) = x\}$. The *intrinsic*
n-skeleton of P is $P^n = \{x \in P \mid I(x, P) \leq n\}$, and the intrinsic n-stratum of P is $P[n] = P^{(n)} - \overline{P}^{(n-1)}$.

For an example, consider again the polyhedron P which is
the wedge of the 2-simplex σ and the 1-simplex τ at the
vertex v. Then $P^{(2)} = P$, $P^{(1)} = P - (\text{int } \sigma) = (\text{bd } \sigma) \cup \tau$,

and $P^{(0)}$ = bd τ. Therefore $P[2]$ = int σ, $P[1]$ = (bd σ - v) $\cup$ (int τ), and $P[0]$ = bd τ.

Three properties of this example should be noted because they hold for the above defined stratification on any polyhedron P. First, each $P^{(i)}$ is a closed subpolyhedron of P. Secondly, each $P[i]$ is a topological n-manifold. Less obvious, but crucial to the study of cellular sets in polyhedra, is the fact that each stratum of P is an isotopy class of P. That is, for each x, y $\in$ P, there is an isotopy of P taking x to y if and only if x and y lie in the same component of the same stratum of P.

We now point out an example where the n-stratum of P does not inherit a P.L. manifold structure from P. Let H^3 be a homology 3-sphere such that $\pi_1(H^3) \neq 0$. If $P = \sum^2 H^3$, then $P = P[5]$, but $P[5]$ does not inherit a P.L. manifold structure since it contains a wildly embedded 1-dimensional subcomplex, the suspension circle.

3. CELLULAR SETS

The key to the intimate connection between the concept of a cellular subset X of a polyhedron P and the given stratification of P is contained in the following proposition.

PROPOSITION 3.1: [8] Let X be a cellular subset of P and $g_t : P \to P$ a pseudoisotopy shrinking precisely X. Then if $P[\ell]$ is the lowest dimensional stratum that X intersects, X intersects only one component A of $P[\ell]$ and $g_1(X) \in A$.

The proof of this proposition and the next rely heavily on the fact that each component of each stratum of P is an isotopy class of P.

PROPOSITION 3.2:   Let   X   be a compact subset of   P   and P[ℓ]   lowest dimensional stratum of   P   that   X   intersects. Then   X   is cellular in   P   if and only if $X \cap P^{(i)}$ is cellular in $P^{(i)}$ for $i \geq \ell$.

Proof.   If   X   is cellular in   P,   then there is a pseudo-isotopy $g_t : P \to P$ shrinking precisely   X.   Let $g_t^i = g_t | P^{(i)}$. Then g_t^i is a pseudoisotopy of $P^{(i)}$ which has $X \cap P^{(i)}$ as the only nondegenerate point preimage of g_1^i.

If $X \cap P^{(i)}$ is cellular in $P^{(i)}$ for each $i \geq \ell$ and $n = \dim P$, $X = X \cap P^{(n)}$ is cellular in $P^{(n)} = P$.

The last result of this section is the generalization of the similar characterizations of cellular sets in manifolds.

THEOREM 3.3:   [8]   Let   X   be a compact subset of a poly-hedron   P.   Then the following are equivalent:

(1)   X   is cellular in   P

(2) $\pi : P \to P/X$ is approximable by homeomorphisms

(3) $X = \cap N_i$,   where N_1, N_2, $\cdots$ are each homeomorphic to $\mathbb{R}^\ell \times cL$ and $\overline{N}_{i+1} \subset N_i$.

4.   CELLULAR MAPS

A proper surjection $f : P \to Q$ is a *cellular map* if for each $y \in Q$,   the set $f^{-1}(y)$ is cellular in   P.   Except for Theorem 4.3, we will assume as a hypothesis for the theorems in this section that $f : P \to Q$ is a cellular map.

Some of the first results on the approximability of cellu-lar maps by homeomorphism are those of Moore   (n = 2),   Armen-trout   (n = 3),   and Siebenmann   (n $\neq$ 4).   We combine these re-sults into one theorem.

THEOREM 4.1:   [11, 2, 12]   If   P   and   Q   are both topological n-manifolds,   n ≠ 4,   then   f   is approximable by
homeomorphisms.

Handel was able to use this theorem to make an advance in
the direction of general cellular maps.

THEOREM 4.2:   [7] If for each $y \in Q[n]$, $f^{-1}(y)$ is a
cellular subset of   P[n]   and   P[4] = Q[4] = ∅,   then   f   is
approximable by homeomorphisms.

The first generalization of Theorem 4.1 which weakened the
condition on the polyhedra   P   and   Q   rather than the point
preimages of   f   concerned the situation where either   P   or
Q   is a generalized n-manifold. A polyhedron   P   is a generalized n-manifold if for each $x \in P$, $H_*(P, P - x) \cong$
$\cong H_*(R^n, R^n - 0)$.   Cannon has shown that polyhedral generalized
n-manifolds come very close to being topological n-manifolds.

THEOREM 4.3:   [5] A polyhedral generalized n-manifold   P
is locally an n-manifold except possibly at the vertices of
P. If $n \leq 3$,   or $n \geq 5$ and $\pi_1(\ell k(v, P)) = 0$,   then   P   is
locally an n-manifold at the vertex   v   of   P.

Using this result and Theorem 4.1, we have the following.

THEOREM 4.4:   [8] If   P   or   Q   is a generalized n-manifold,   n ≠ 4,   then   f   is approximable by homeomorphisms.

The next theorem is not an approximation theorem, but it
does imply that if $f : P \to Q$ is cellular, there is a   1 - 1
correspondence between the components of the n-strata of   P
and   Q   provided   Q[4] = ∅.   This theorem is crucial to the
proof of the following results.

THEOREM 4.5:   [8, 9] Suppose that $Q[4] = \emptyset$. Then for each component B of each stratum $Q[n]$ of Q, there is a component A of $P[n]$ such that $f|\overline{A} : \overline{A} \to \overline{B}$ is a cellular map. Also, $f|P^{(i)} : P^{(i)} \to Q^{(i)}$ is a cellular map.

COROLLARY 4.6:   Suppose that $Q[4] = \emptyset$ and for each $y \in Q[n]$ the set $f^{-1}(y)$ lies in some stratum of P. Then f is approximable by homeomorphisms.

Proof. It follows from the previous theorem that if $y \in B$, where B is a component of the n-stratum of Q, then $f^{-1}(y) \subset \overline{A}$ for some component A of $P[n]$. If $f^{-1}(y) \not\subset A$, then there is a component A' of $P[k]$, $k < n$, such that $f^{-1}(y) \subset A'$. But then $f(P^{(n-1)}) \not\subset Q^{(n-1)}$, contradicting the above theorem. Hence $f^{-1}(y) \subset A$.

The previous theorem also implies that $P[4] = \emptyset$. We may now apply Theorem 4.2 to get the desired result.

THEOREM 4.7:   [8] If either P or Q is an n-manifold with boundary, $n \neq 4, 5$, then f is approximable by homeomorphisms.

THEOREM 4.8:   [9] If $\dim P \leq 3$ or $\dim Q \leq 3$, then f is approximable by homeomorphisms.

Question 4.9. Is there a cellular map between polyhedra which is not approximable by homeomorphisms?

5. CELLULAR DECOMPOSITIONS OF POLYHEDRA

The natural next step is to look at a cellular upper semicontinuous decomposition (u.s.c.d.) of a polyhedron P. One cannot show that an arbitrary cellular u.s.c.d. of a polyhedron

P yields the same space. For example, let G be the cellular u.s.c.d. of the 3-cell B^3 into points and a cantor set
of arcs, each arc having one end point in the boundary of B^3,
such that B^3/G is the Alexander crumpled cube. Furthermore,
$B^3/G \times \mathbb{R}^1$ is not homeomorphic to $B^3 \times \mathbb{R}^1$. Thus the open
question of whether each cellular u.s.c.d. G of an n-mani-
fold M^n has the property that $M^n/G \times \mathbb{R}^1 \cong M^n \times \mathbb{R}^1$ does not
carry over to the realm of cellular u.s.c.d. of polyhedra.

It can easily be shown that the question of being able to
raise dimension with a cellular u.s.c.d. of a polyhedron can
be answered if one can prove that a cellular u.s.c.d. of a
manifold does not raise dimension. If we define $G[n] =$
$= \{g \cap P[n] | g \in G, \ g \cap P[n] \neq \emptyset, \ \text{and} \ g \cap P^{(n-1)} = \emptyset\}$, then
it follows from Proposition 3.2 that $G[n]$ is a cellular
u.s.c.d. of the n-manifold $\tilde{P}[n] = P[n] - \cup \{g \in G | g \cap P^{(n-1)}$
$\neq \emptyset\}$. Hence if $\dim \tilde{P}[n]/G[n] = n$, then $\dim P/G = \dim P$.

Question 5.1. Under what condition is the decomposition
space P/G homeomorphic to P if P/G is a CS set? (See
[7] for definition of CS set.)

Question 5.2. Is there a disjoint disks property for a
cellular u.s.c.d. of polyhedra?

6. POLYHEDRAL CELLULARITY CRITERION
In 1964, McMillan published the cellularity criterion for cell-
like sets in the interior of a P.L. n-manifold, $n > 5$.

THEOREM 6.1: [10] Let X be a cell-like set in the interior of a P.L. n-manifold M^n, $n \geq 5$. A necessary and sufficient condition that X be the intersection of a sequence

of P.L. n-cells $\{F_i\}$ where $F_{i+1} \subset$ int $F_i \subset M^n$ is that the following property hold:

For each open set U containing X, there is an open set V such that $X \subset V \subset U$ and each loop in $V - X$ is null homotopic in $U - X$.

When an attempt was made to generalize this to a cellularity criterion for a certain class of subsets of polyhedra, it became clear that one could not consider all cell-like subsets of polyhedra as candidates to which a cellularity criterion could be applied. For example, let P be an n-cell B^n. Then P is a cell-like subset of itself, and for each neighborhood U of P in P, $U - P = \emptyset$ so any condition on $U - P$ is trivially satisfied. A more restrictive class of subsets of P is needed.

A homotopy $h_t : Z \to P$ is *stratum respecting* if for each $z \in Z$, $I(h_t(z), P) \geq I(h_s(z), P)$ for $s \geq t$. A *rooted cell-like* set X in P is a compact subset of P such that for each neighborhood U of X, there is a neighborhood V of X and a stratum respecting contraction $h_t : V \to U$ such that h_0 is the natural inclusion.

The only compact polyhedron P which is a rooted cell-like subset of itself is a point, and the previous problem is eliminated. Also, using part (3) of Theorem 3.3, it is easy to see that each cellular subset of P is also a rooted cell-like set in P. One drawback is that X being a rooted cell-like subset of the polyhedron P is a property of the embedding of X in P, and not a property of X itself.

An obvious attempt to generalize McMillan's criterion would be to require that for each neighborhood U of a rooted cell-like set X in P, there is a neighborhood V of X such that each loop in $(V \cap P[i]) - X$ is null homotopic in $(U \cap P[i]) - X$. However, this is not the case for some cellular subsets of polyhedra. Let P be the wedge of the open cone on $S^1 \times B^3$, and the 1-cell B^1, with the wedge point taken at the cone point and a vertex of B^1. If we denote the open cone on $S^1 \times B^3$ by $\dfrac{[0,1) \times S^1 \times B^3}{\{0\} \times S^1 \times B^3}$, the cellular set X is $\dfrac{[0,\frac{1}{2}] \times S^1 \times B^3}{\{0\} \times S^1 \times B^3}$. Given a neighborhood U of X, each neighborhood V of X with $X \subset V \subset U$ will contain a loop, generated by the S^1 factor, which is not null-homotopic in U - X.

We now state the polyhedral cellularity criterion.

Polyhedral Cellularity Criterion: A compact subset X of a polyhedron P satisfies the polyhedral cellularity criterion (PCC) if for each open set $U \supset X$, there is an open set $V \supset X$ such that $X \subset V \subset U$ and for every stratum P[i], $i \geq 3$, each singular k-cell $D^k \subset U \cap P[i]$ with $bdD^k \subset (V - X) \cap P[i]$ is homotopic rel bdD^k in $U \cap P[i]$ to a singular k-cell $B^k \subset (U - X) \cap P[i]$ for k = 1, 2.

THEOREM 6.1: [8] Let P be a polyhedron such that each P[i] inherits a P.L. i-manifold structure from P, and let X be a rooted cell-like set in P which intersects no stratum of dimension less than 5. Then X is cellular in P if and only if X satisfies the PCC.

The criterion would not be of much interest if all rooted cell-like sets were also cellular. However, there is a simple example of a rooted cell-like set which is not cellular. Again, let G be the u.s.c.d. of B^3 such that B^3/G is the Alexander crumpled cube. The rooted cell-like set X is then $(\cup\{g \in G | g$ not a point$\}) \cup A$, where A is an arc embedded in bdB^3 such that A intersects each nondegenerate element of G. Since $B^3/X \not\approx B^3$, X is not cellular, but X is a rooted cell-like set.   Higher dimension examples can be similarly constructed.

Finally, one should note that in the case where   P   is a P.L. n-manifold, the PCC reduces to McMillan's cellularity criterion.

Question 6.2.   Can the hypothesis concerning the ability to move singular 1-cells be eliminated from the PCC?

Question 6.3.   Is there a cellularity criterion for cell-like sets in lower dimensional polyhedra?

REFERENCES

[1]   Aiken, E., "Manifold phenomena in the theory of poly-
        hedra." Trans. Amer. Math. Soc., 143 (1969), 413-473.
[2]   Armentrout, S., "Cellular decompositions of 3-manifolds
        that yield 3-manifolds." Memoir 7, Amer. Math. Soc.
        (1971).
[3]   Armstrong, M., "Transversality for polyhedra." Ann.
        Math. 86 (1967), 172-191.
[4]   Brown, M., "A proof of the generalized Schoenflies
        theorem." Bull. Amer. Math. Soc. 66 (1960), 74-76.
[5]   Cannon, J. W., "The Recognition problem: what is a
        topological manifold?" Bull. Amer. Math. Soc. 84
        (1978), 832-866.
[6]   ------, "Shrinking cell-like decompositions of mani-
        folds: codimension three." Ann. Math. 110 (1979),
        83-112.
[7]   Handel, M., "Approximating stratum preserving CE maps
        between SC sets by stratum preserving homeomorphisms,"

in Geometric Topology, (L.C. Glaser and T.B. Rushing, eds.) Lecture Notes in Math. 438, Springer-Verlag, New York, 1975, 245-250.

[8]   Henderson, J., "Cellularity in polyhedra, Topology and its Applications," to appear.

[9]   ------, "Approximating cellular maps between low dimensional polyhedra," preprint.

[10] McMillan, D. R., Jr., "A criterion for cellularity in a manifold." Ann. Math. 79 (1964), 327-337.

[11] Moore, R. L., "Concerning upper semicontinuous collections of continua." Trans. Amer. Math. Soc. 27 (1925), 416-428.

[12] Siebenmann, L.C., "Approximating cellular maps by homeomorphisms." Topology 11 (1972), 271-294.

[13] ------, "Deformations of homeomorphisms on stratified sets." Comm. Math. Helv. 74 (1972), 123-163.

Raising the Dimension of 0-Dimensional Decompositions of E^3

Louis F. McAuley and Edythe P. Woodruff

1. <u>Introduction</u>. We shall be concerned with upper semicontinuous (usc) decompositions G of E^3 using standard definitions (Whyburn [3]). As usual, H will denote the collection of non-degenerate elements of G and $P : E^3 \Rightarrow E^3/G$ will
denote the quotient (or projection) map onto the decomposition
space E^3/G. A decomposition is called point-like if, for
each $g \in H$, $E^3 - g$ is homeomorphic to $E^3 - $ (a point).

In [2] we have developed techniques to answer a question
asked by Robert Daverman. Our solution involves altering a
0-dimensional decomposition so that the dimension of $P(H)$ is
raised. Since this dimension-raising technique is of interest
in its own right, in this paper we will generalize it and
raise some further related questions. For the purposes here,
an alternate description of the example in [2] is necessary.
In Section 2 we give that alternate description. Sections 3
and 4 are the generalizations and questions.

2. <u>Question</u>. (Daverman) Does there exist a point-like usc
decomposition G of E^3 such that
  (1) $P(H)$ has dimension 1;
  (2) E^3/G is not homeomorphic to E^3;
  (3) for any $H' \subseteq H$ satisfying
      (a) the corresponding decomposition G' is usc, and
      (b) $P(H')$ has dimension 0 in E^3/G;
      it is true that E^3/G' is homeomorphic to E^3;
  (4) each element of H is an arc;
  (5) H^* is a 2-cell; and
  (6) H is a continuous collection?

(With conditions (4)-(6) omitted, this is a question asked by Michael Starbird.)

To construct an example with these properties we change a 0-dimensional decomposition G_0 (i.e., a decomposition G_0 such that $P(H_0)$ is 0-dimensional in E^3/G_0) into a 1-dimensional decomposition G_1. If E^3/G_0 is not homeomorphic to E^3, then this E^3/G_1 fails to be homeomorphic to E^3. Not all 0-dimensional decompositions altered by our techniques result in decompositions satisfying (3). Further modifications are required to make the decomposition satisfy (4)-(6); not all those decompositions satisfying (1)-(3) lend themselves to the further modifications that we use. However, Bing's "Dog Bone" decomposition [1] can be altered to yield the desired example.

A description of the example. The construction of the "Dog Bone" decomposition can be described as the iteration of sets A_i of solid double tori, or "dog bones," embedded as indicated in Figure 1. The set H is the set of components of $\cap_{i=1}^{\infty} A_i$.

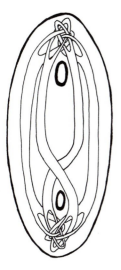

Fig. 1.

154   *Louis F. McAuley and Edythe P. Woodruff*

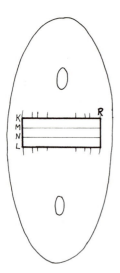

Fig. 2.

The "dog bones" can be so placed that each element of H intersects the rectangle R indicated in Figure 2 in a vertical
line segment. Let K denote the upper edge of R and L
denote the lower edge. Let M and N be the line segments
1/3 and 2/3 of the distance from K to L, as indicated.

We have placed H so that $H^* \cap K$ is the Cantor set C_K
described in the following conventional way using base 3 numbers. To each point of C_K there corresponds in the ternary
notation a number $.a_1a_2a_3\cdots$ where for each i, $a_i = 0$ or
2. Endpoints of K are 0 and $.222\cdots = 1$. Of course,
$H^* \cap L$ is the corresponding Cantor set C_L.

Now, in M and N use base 2 notation to denote each
point and let the endpoints be 0 and $.111\cdots = 1.0$. Let
$f : C_A \Rightarrow M$ be the map taking $.a_1a_2a_3\cdots$ to $.m_1m_2m_3\cdots$
where $m_i = a_i$ if $a_i = 0$, and $m_i = 1$ if $a_i = 2$. This
map is onto M.

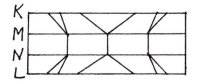

Fig. 3.

Connect each point in C_K with its image in M by a line segment. Using the canonical projection from M to N, connect corresponding points in M and N by line segments. Connect by line segments points in N with the corresponding points in the Cantor set $C_L \subset L$. Call the union of the points in all these line segments J^*. See Figure 3.

Replace $R \cap H^*$ by J^*. Let H_1 be the non-degenerate elements of a new decomposition G_1 and define H_1 by $H_1^* = (H^* - R) \cup J^*$ and each element of H_1 contains exactly one point in M and is the image of one or two elements of H. Notice that a countable number of elements of H_1 are trees which are not arcs and the other elements of H_1 are arcs.

We claim that the decomposition G_1 satisfies Starbird's question.

Observe that $H' = \{$the connected subset of g bounded by points in M and $N | g \in H_1\}$ satisfies all conditions except (2). As a working step, we use $\hat{H} = \{$the closure of the subset of g above $M | g \in H_1\}$. In [2] it is proved that $E^3/\hat{G}$ is homeomorphic to E^3. The proof involves the use of several successive decompositions and their corresponding natural projections. One of these takes that portion of each element which lies above K to a point and distorts the line segment M so that the set of V-shaped arcs between K and M in Figure 3 becomes a null sequence of arcs. Another decomposition takes each arc in this null sequence to a point. In the process, M becomes a wild arc. For full details, please see [2]. Let $\hat{P} : E^3 \Rightarrow E^3/\hat{G}$ be the projection map corresponding to the succession of decompositions and taking each element of $\hat{H}$ to a point.

Let $\hat{\hat{H}}$ = {the closure of the subset of $\hat{P}(g)$ below $\hat{P}(N)$ | $g \in H_1\}$. Then $(E^3/\hat{G})/\hat{\hat{G}}$ is homeomorphic to E^3. In this decomposition space the collection $\{\hat{\hat{P}}(\hat{P}(g)) | g \in G\}$ satisfies Daverman's question.

Comments. Daverman also asked whether a further condition that (7) the 2-cell H^* is cellular, could be satisfied. We do not know.

Details of another description of this example and proofs are included in a paper submitted for publication in the Transactions.

The example constructed in this paper may be related to the question of whether there exists a point-like decomposition of E^3 such that $E^3/G \times E^1$ is not homeomorphic to E^4.

3. An answer to a higher-dimensional problem similar to Daverman's question. There is a point-like usc decomposition G of E^3 such that

    (1) P(H) has dimension 2;

    (2) E^3/G is not homeomorphic to E^3;

(3)   for any H' ⊆ H satisfying

    (a)   the corresponding decomposition G' is usc, and

    (b)   P(H') has dimension less than 2 in E^3/G',

    it is true that E^3/G' is homeomorphic to E^3;

(4)   each element of H is an arc;

(5) H^* is a 3-cell; and

(6)   H is a continuous collection.

This example can be realized by methods very similar to those described in Section 2.

In the first dog bone A_1 place the components of A_2, as shown in Figure 4, so that they intersect the rectangular solid B in vertical line segments. Let K_B denote the upper surface of B.   Let $K_B = X \times Y$.   Iterate the placement shown in Figure 4 in subsequent dog bones, so that $H^* \cap K_B$ is $C_X \times C_Y$, where $C_X \subseteq X$ and $C_Y \subseteq Y$, and each of C_X and C_Y is a Cantor set.

Let M_B, N_B, and L_B denote surfaces in B analogous to lines in R.   Let $f_B : C_X \times C_Y \Rightarrow M_B$ be a map analogous to f.

It should now be obvious that this example can be completed in a manner analogous to the one in Section 2.   The 3-cell containing H^* will have wild upper and lower surfaces.

4. Generalizations and questions. Our techniques involve "sectioning" the collection H of non-degenerate elements of an usc decomposition G of E^3.   Consider the following definitions.

A closed set S contained in $\bar{H}^*$ is said to be a *section* of H if and only if S ∩ g is a point s_g for each g ∈ H. A set T *sections* H if and only if $T \cap \bar{H}^*$ is a section of H.   Suppose that S is a section of H and that A is a subset of E^3 such that there is an embedding θ of S × A into E^3.   The set S × A is said to be a *section product* of

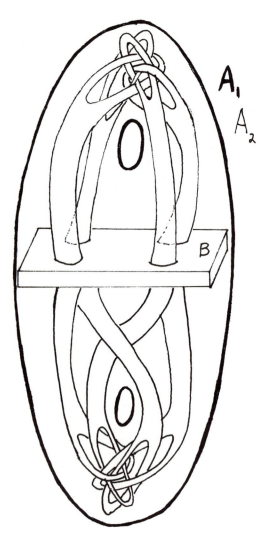

Fig. 4.

H if and only if, for some embedding θ and for each $g \in H$,
it is true that $S \cap g = s_g$ and $\theta(s_g,A) \subset g$.

In the example described in Section 2, C_K is a section
of H, the segment K sections H, and for an arc A there
is a section product with A. The original collection H has
a section product contained in the disk R; the altered collection H_1 has a section product which is the disk R.

As above, G denotes an usc decomposition of E^3 and H
is the collection of non-degenerate elements of H. Note that
these ideas extend to E^n (and S^n).

Question. For what collections H and sets A is it true
that H has a section product with A?

Question. What decompositions can be sectioned by an arc?
There are collections H for which $P(\bar{H})$ has dimension 0
and H cannot be sectioned by an arc.

Question. If a collection H is sectioned by an arc, then
is there a tame arc which sections H?

Question. If H is a closed 0-dimensional decomposition
(i.e., $P(\bar{H})$ is 0-dimensional) and H is a continuous collection, then can H be sectioned by an arc? If $P(H)$ is not
required to be closed, the answer is no.

The next theorem is for use in the definition which follows
it.

THEOREM. Suppose that G is an usc decomposition of E^3
and that H is the collection of non-degenerate elements of
G. Furthermore, there exists a section S for H such that
  (1) S is contained in a plane Q.
  (2) $Q \cap H^* = S$, which is a Cantor set, and
  (3) there is a section product $S \times A$ of H where A is
       an arc; and
  (4) it is possible to choose the embedding θ so that for
       some $a \in \text{Int } A$, the section $S = \theta(S,a)$.

Then, for any non-empty compact set K in Q, there exists
a mapping f of E^3 onto itself such that

(a)   f(Q) = Q,

(b)   f(S) = K, and

(c)   f is a homeomorphism on E^3 - Q.

Proof. Let $\{N_i\}_{i \in \omega}$ be a sequence of nested closed
neighborhoods of Q such that

(i) N_i is bounded by planes R_i and T_i parallel to Q

(ii) $N_{i+1} \subset \text{Int } N_i$, and

(iii) $\cap_{i \in \omega} N_i = Q$.

The Cantor set S can be considered the intersection of
closed manifolds $\{M_i\}_{i \in \omega}$ such that

(i) $M_{i+1} \subset \text{Int } M_i$,

(ii)   each component of M_i is a 2-cell in Q, and

(iii)   each component of M_i contains 4 components of M_{i+1}.

Denote a component of M_i by $M(n_1,n_2,\cdots,n_j,\cdots,n_i)$,
where each $n_j = 1,2,3$, or 4 and

$$M(n_1,n_2,\cdots,n_j) \supset M(n_1,n_2,\cdots,n_j,\cdots,n_i) .$$

Let B_0 be a square in Q such that $K \subset B_0$. Let B_i be
the collection of 4^i squares formed from B_0 by cutting on
2^i equally spaced lines parallel and 2^i equally spaced lines
perpendicular to one side of B_0. Denote each square in B_i
by $B(n_1,n_2,\cdots,n_j,\cdots,n_i)$ in a manner analogous to notation
for components of M_i. Let $C(n_1,n_2,\cdots,n_j)$ be the rectangular solid with faces in Bd N_1 and intersecting Q in
$B(n_1,n_2,\cdots,n_j)$.

First, we can consider a sequence of homeomorphisms $\{\phi_i\}_{i \in \omega}$
such that

(i) $\phi_i = \phi_{i-1}$ off $N_{i-1} \cap C(n_1)$, and

(ii) ϕ_i takes each $N_i \cap M(n_1,n_2,\cdots,n_j)$ into the corresponding $N_i \cap C(n_1,n_2,\cdots,n_j)$.

The map $\phi = \lim_{i \to \infty} \phi_i$ has the properties

(a) $\phi(Q) = Q$,

(b) $\phi(S) = B_0$, and

(c) ϕ is a homeomorphism on $E^3 - Q$.

In order to change the image of S in (b) to K we modify ϕ_i to f_i by changing only condition (ii). Now consider, for each $B(n_1,n_2,\cdots,n_j)$, whether it intersects K. If so, then let $f_i|N_i \cap M(n_1,n_2,\cdots,n_j) = \phi_i|N_i \cap M(n_1,n_2,\cdots,n_j)$. If not, let $B(n_1,n_2,\cdots,n_j')$ be a square which does not intersect K. (There must be such an n_j'.) Using this n_j', require that f_i be a homeomorphism taking $N_i \cap M(n_1,n_2,\cdots,n_j)$ into $N_i \cap C(n_1,n_2,\cdots,n_j')$; and then for f_k where $k > i$ make f_k take $N_k \cap M(n_1,n_2,\cdots,n_j,\cdots,n_k)$ into this $N_k \cap C(n_1,n_2,\cdots,n_j')$ and let $f_k|N_k \cap C(n_1,n_2,\cdots,n_j,\cdots,n_k) = f_{k-1}|N_k \cap C(n_1,n_2,\cdots,n_j,\cdots,n_k)$.

Let f_k for $k > i$ take all components of

$$N_k \cap M(n_1,n_2,\cdots,n_j,\cdots,n_k) \quad \text{into} \quad N_i \cap M(n_1,n_2,\cdots,n_j',\cdots,n_k).$$

Now $f = \lim_{i \to \infty} f_i$ satisfies the theorem.

(Notice that if one wishes one can define this map more carefully and obtain one which is no worse than 4 to 1 at any point of K.)

For each $k \in K$, let $g_k = \cup \{f(h) | h \in H$ and $k \in f(h)\}$. Let $H_1 = \{g_k | k \in K\}$. The decomposition G_1 of E^3 whose collection of non-degenerate elements is H_1 is upper semi-continuous. We call G_1 a (G,K,f)-*induced decomposition of* E^3. Clearly, $P_1(K)$ is homeomorphic to K where $P_1 : E^3 \Rightarrow E^3/G$ is the projection mapping.

Question. For what (G,K,f)-induced decompositions is it true that if $H_1' \subset H_1$ and (i) the associated G_1' is usc and (ii) $P_1(H_1')$ is 0-dimensional in E^3/G_1', then E^3/G_1' is homeomorphic to E^3?

ACKNOWLEDGMENT

This research was supported in part by the National Science Foundation under Grant MCS-79 09542.

REFERENCES

[1] R. H. Bing, "A decomposition of E^3 into points and tame arcs such that the decomposition space is topologically different from E^3," Annals of Math. 65 (1957), 484-500.

[2] L. F. McAuley and E. P. Woodruff, "Certain point-like decompositions of E^3 with 1-dimensional images of non-degenerate elements," submitted to Transactions of the American Mathematical Society.

[3] G. T. Whyburn, Analytic Topology, AMS Colloquium Publications 28 (1942).

Curves Isotopic to Tame Curves

M. Brin

This paper concerns the following: Is Bing's simple closed
curve that pierces no disk isotopic to a tame simple closed
curve?

We give a description of Bing's curve J_1 and then describe
another curve J_2 which at each point is locally equivalent
to J_1 but which is isotopic to a tame simple closed curve.
Thus any demonstration that Bing's curve is not isotopic to a
tame curve cannot be based on local considerations only. The
picture we give defining J_1 is not identical to the picture
used to define Bing's curve [1], but it is easy to show that
there exists a homeomorphism from E^3 to itself carrying
Bing's curve onto J_1. The curve J_2 has also been known for
some time to J. W. Cannon.

The isotopy connecting J_2 to a tame simple closed curve
gives a tame curve at every level but one. As an added
attraction we supply a description of another curve J_3 which
pierces no disk and which is isotopic to a tame simple closed
curve via an isotopy in which every level but one produces a
simple closed curve that pierces no disk.

Bing's curve can be defined using Figure 1. Solid torus
T_1 is the union of the 3-cells $B_{1,0}$ through $B_{1,5}$. Solid
torus T_2 is the union of the 3-cells $B_{2,i}$. Only five of
the cells $B_{2,i}$ are shown. The curve in Figure 1 indicates
the centerline of the rest of T_2. A solid torus T_3 is contained in T_2 and is the union of 3-cells $B_{3,j}$. The centerline of T_3 is shown for that portion of T_3 in $B_{2,3}$. Each
$B_{2,i}$ contains 3-cells $B_{3,j}$ arranged in the same manner as
various $B_{2,k}$ are arranged in $B_{1,0}$. Similarly solid tori

$T_4 \subseteq T_3$, $T_5 \subseteq T_4$, ... are defined. Bing's curve J_1 is the intersection of the T_i.

The universal cover of T_1 is shown in Figure 2. We assume that exactly k $B_{2,i}$ are contained in each $B_{1,j}$. The picture in Figure 1 is recovered by identifying each $B_{1,j}$ with $B_{1,j+6}$. We will describe a second curve by deriving a different picture from Figure 2. From Figure 1 remove all $B_{1,j}$ where $j < 0$. Also ignore all $B_{2,i}$ where $i < 0$. This second admonition says to ignore those $B_{2,i}$ that form a neighborhood of the arc A in $B_{1,0}$. Retain all $B_{r,s}$ with $r \geq 3$ so that $B_{r,s}$ lies in some $B_{2,i}$ for $i \geq 0$. Note that the 3-cells $B_{2,i}$ for $i \geq 0$ follow a ray that can be obtained from a straight ray by an isotopy (an infinite number of crochet operations). The picture in Figure 3 is obtained by identifying each $B_{1,j+6}$ with $B_{2,j}$ for $j \geq 0$. The curve J_2 is the set of points contained in an infinite number of the $B_{1,j}$. It seems likely that J_1 and J_2 are everywhere locally equivalently embedded (make up your own definition and supply your own proof). It is also clear that J_2 can be obtained from a tame simple closed curve by an isotopy (an infinite number of crochet operations). This isotopy, defined on $S^1 \times [0,1]$, takes each $S^1 \times \{t\}$, $0 \leq t < 1$, to a tame curve and $S^1 \times \{1\}$ to a simple closed curve that pierces no disk.

The curve J_3 is a modification of the curve J_2. Construct a picture similar to Figure 3 except that the cells $B_{2,j}$ for $j \geq 0$ should follow the ray pictured in Figure 4 rather than the ray of Figure 2. This new ray is obtained from a straight ray by an isotopy (an infinite number of twisted crochets). The crochets are peculiar in that they can be undone by pulling at the free end of the ray in C_0. Define J_3 using Figure 4 as J_2 was defined using Figure 2. There is

now an isotopy, similar to that connecting J_2 to a tame
simple closed curve, defined on $S^1 \times [0,1]$ so that each
$S^1 \times \{t\}$, $0 \le t < 1$, is carried to a tame curve and so that
$S^1 \times \{1\}$ is carried to J_3, a simple closed curve that
pierces no disk. Now J_3 can be isotopically untangled "in
reverse" by grabbing hold (say somewhere near C_6) and pulling
out the large loops (crochets) that occur in C_1, C_2, C_3, C_4
and C_5. A tighter hold can be obtained and the loops (crochets)
in C_6, C_7, $C_8, \cdots$ can be undone. This gives an isotopy defined on $S^1 \times [1,2]$ so that each $S^1 \times \{t\}$, $1 \le t < 2$, is
carried to a simple closed curve that pierces no disk and so
that $S^1 \times \{2\}$ is carried to a tame curve.

Exercises: Define a simple closed curve J_4 with Figure 3
as the base picture and using the reverse of untwisted crochet-
ing (the mirror image of Figure 2). Show that J_4 is isotopic
to a tame curve and wonder if every isotopy of J_4 to a tame
curve must contain a half open interval of curves that pierce
no disk.

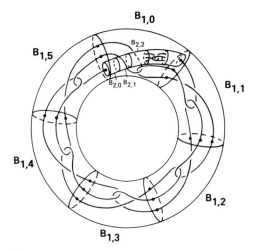

Figure 1

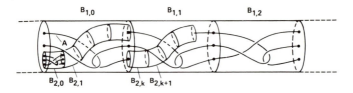

Figure 2

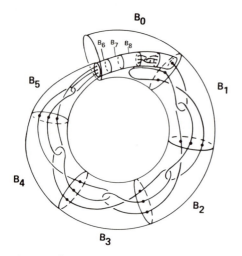

Figure 3

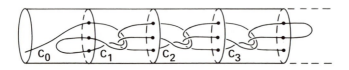

Figure 4

REFERENCE

[1] Bing, R. H., <u>J. Math. Pures and Appl.</u> (9)35 (1956), 339.

Symplectic Maps on Open Cells:
A Fixed Point Theorem

M. R. Colvin and K. Morrison

INTRODUCTION

Let B be a 2n-dimensional open cell in R^{2n}. A symplectic structure on B is the geometrical structure induced by a differentiable exterior 2-form ω defined on B, satisfying the conditions: (1) the form ω is closed, i.e., $d\omega = 0$; and (2) ω is everywhere of maximal rank, i.e., the 2^n-form ω^n (nth exterior power of ω) is everywhere different from zero or equivalently the bundle mapping $\tilde{\omega} : TB \to T^*B$, defined by $\tilde{\omega}(x)(y) = \omega(x,y)$ is an isomorphism. [A&M] The classical example of a symplectic structure on the cell B is the structure given by picking coordinates for R^{2n}, say $(x_1, x_2, \cdots, x_{2n})$, and defining

$$\omega = \sum_{j=1}^{n} dx_{2j-1} \wedge dx_{2j} \quad .$$

A differentiable mapping ψ of B into itself is called symplectic if it preserves the two form ω, i.e., if $\psi^*\omega = \omega$. [A&M] The purpose of this study is to establish a fixed point theorem for symplectic maps defined on the open 2n-cell B that generalize corresponding results of Weinstein [W] and Moser [M] for compact symplectic manifolds. The original results for the open two-disk were initiated by Bourgin, [B].

Throughout, our notation and terminology will conform to that of [C&M] -- to which we refer for more general results in the setting of open symplectic manifolds. This is an expanded version of an address given by the first author during the Summer Topology Conference held at Austin, Texas, from June 1, 1980 through July 15, 1980.

PRELIMINARIES

We shall denote the collection of smooth symplectic diffeo-morphisms defined on the symplectic manifold (M, ω) by $\text{Diff}(M, \omega)$. The collection of closed one forms will be denoted by $Z^1(M)$. Both function spaces will have the C^1-fine topology. (See [H] for a discussion of the C^1-fine topology.)

If (M, ω) is a symplectic manifold, there is a natural symplectic two-form ω_0 on T^*M. Using the isomorphism $\tilde{\omega} : TM \to T^*M$ induced by ω, the two-form ω_0 "pulls back" to induce a natural two-form ω_1 on TM. There is also a natural symplectic structure induced on $M \times M$ by $\omega_2 = \Pi_1^* \omega - \Pi_2^* \omega$, where $\Pi_i : M \times M \to M$, $i = 1, 2$, are the natural projections. If ψ is a symplectic diffeomorphism of (M, ω), then the graph of ψ is a Lagrangian submanifold of $(M \times M, \omega_2)$. In [W], Weinstein describes a symplectic diffeomorphism F of a neighborhood of the diagonal in $(M \times M, \omega_2)$ onto a neighborhood of the zero section in (TM, ω_1) which sends the graphs of symplectic maps to graphs of symplectic vector fields. Using the symplectic diffeo-morphism F, Weinstein constructs:

THEOREM (1.1): [W, Theorem 7.2]. If (M, ω) is a symplec-tic manifold, then there are C^1-fine neighborhoods, U, of the identity diffeomorphism of M in $\text{Diff}(M, \omega)$, and W, of the zero one-form in $Z^1(M)$ and a homeomorphism $V : U \to W$. Moreover, if $\psi \in U$, then a point $x \in M$ is a fixed point of ψ, i.e., $\psi(x) = x$, if and only if $V(\psi)(x) = 0$.

Assume M is a differentiable manifold, with or without a symplectic structure. Also, assume η is a closed one-form defined on M. We denote by $C(\eta)$ the number of zeros of η

and by $C(M) = \inf\{C(\eta) \mid \eta \in Z^1(M)\}$. If M is simply connected and $\eta \in Z^1(M)$,   then there exist a real valued function   f   defined on   M,   so that $df = \eta$. In the case where
M   is also assumed compact, then   f   has a maximum and a
minimum. It follows, that any two-form defined on the compact, simply connected manifold M has at least two zeros
and therefore $C(M) \geq 2$.

THE MAIN THEOREM

In the following, we shall assume N is a fixed point of
S^{2n},   the unit sphere in $\mathbb{R}^{2n+1}$. The 2n-cell, B, can be
embedded diffeomorphically onto $S^{2n} - \{N\}$. Using the differential of the above embedding, one-forms on S^{2n} induce
one-forms on B. The collection of one forms, η, on S^{2n}
which vanish at N, i.e., $\eta(N) = 0$, induce an open neighborhood of the zero form in $Z^1(B)$. It follows that we can
assume in Theorem (1.1) that the neighborhood W is induced
by the above embedding, i.e., $\eta \in W \subseteq Z^1(B)$ if and only if
η extends to $\tilde{\eta} \in Z^1(S^{2n})$ and $\tilde{\eta}(N) = 0$. Since S^{2n} is
compact and simply connected, then $C(S^{2n}) \geq 2$. It follows,
if $\eta \in Z^1(B)$ then the induced one-form, $\tilde{\eta} \in Z^1(S^{2n})$ has
at least two zeros; therefore η has at least one zero and
we have:

LEMMA (2.1): There is an open C^1-fine neighborhood, W,
of the zero form in $Z^1(B)$, so that every one form in W has
at least one zero.

Using the isomorphism, V, given in Theorem (1.1) we can
establish:

THEOREM (2.2): Let B denote the open 2n-dimensional cell in $\mathbb{R}^{2n}$ and ω a symplectic structure on B, then there exists a C^1-fine neighborhood, U, of the identity in Diff(B,ω), so that if $\psi \in U$, then ψ has at least one fixed point.

Proof. Using Theorem (1.1), there are C^1-fine open subsets $U \subseteq$ Diff(B,ω) and $W \subseteq Z^1(B)$ containing the identity and the zero one-form respectively and a homeomorphism $V : U \to W$ so that $V(\psi)(x) = 0$ if and only if $\psi(x) = x$. Using Lemma (2.1) each one-form in ω has at least one zero, hence for $\psi \in U$, ψ has at least one fixed point.

CONCLUSION

In Bourgin's [B] study he shows that every area preserving, orientation preserving homeomorphism of the open two-cell has at least one fixed point. Using recent results by Simon [S] our techniques yield Bourgin's results. However, for higher dimensions, it remains an open question if every symplectic diffeomorphism of an open 2n cell has a fixed point.

For odd dimensional cells one replaces symplectic structures by contact structures. In [C&M], the odd dimensional case is studied and a somewhat weaker result than (2.1) holds. Again, the general result, "Does every contact transformation defined on the 2n + 1-cell, B, equipped with some contact structure, have a fixed point?" remains an open question.

ACKNOWLEDGMENT

The first author gratefully acknowledges partial support for his participation in the conference from the Department of Mathematics, The University of Texas at Austin.

REFERENCES

[A&M]   R. Abraham and J. E. Marsden, Foundations of Mechanics, Benjamin/Cummings, Reading, Mass., 1978.

[B]   D. G. Bourgin, "Homeomorphisms of the open disk," Studia Mathematica 31 (1968), 433-438.

[C&M]   M. R. Colvin and K. Morrison, "A symplectic fixed point theorem on open manifolds," submitted to Trans. Amer. Math. Soc., 1980.

[H]   M. W. Hirsh, Differential Topology, Springer-Verlag, New York, 1976.

[M]   J. Moser, "A fixed point theorem in symplectic geometry," Acta Mathematica 141 (1978), 17-34.

[S]   C. P. Simon, "A bound for the fixed point index of an area preserving map with applications to mechanics," Inventioned Math. 26 (1974), 187-200.

[W]   A. Weinstein, "Lagrangian submanifolds and Hamiltonian systems," Annals of Math. 98 (1973), 377-410.

A Structure Set Analogue of Chapman-Ferry-Quinn Theory

F. T. Farrell and W. C. Hsiang

I. INTRODUCTION AND STATEMENT OF THE RESULT

Let $F \to E \overset{p}{\to} M$ be a locally trivial fibration and let $f : E' \to E$ be a homotopy equivalence. Assume that F is a finite dimensional compact space and both E, M are finite dimensional locally compact spaces. Moreover, M has a fixed metric d. Given $\varepsilon > 0$, we say that f *is an ε-homotopy equivalence* (or *ε-equivalence*) *with respect to* p if there exists such a homotopy inverse g of f for which there are homotopies H from gf to $id_{E'}$ and G from fg to id_E, that satisfy the following

(i) $\forall x' \in E'$ the path $pfH(x' \times I)$ is an ε-neighborhood of every $y \in pfH(x' \times I)$ i.e., $\text{diam } pfH(x' \times I) < \varepsilon$,

(ii) $\forall x \in E$, $\text{diam } pG(x \times I) < \varepsilon$.

Similarly, we define an ε-equivalence of f over a subset of M and an ε-homotopy of a subset of E with respect to the map fp, etc. Let us consider the following natural questions.

Question 1. Under what conditions is the above map f *homotopic to a homeomorphism?*

Question 2. *Assume that both* E, E' *are compact. Under what conditions is the above map* f *simple?*

In [1], Chapman-Ferry proved that if $F = pt$ and M is a closed manifold of dimension ≥ 5 (or M is a manifold with

boundary of dimension ≥ 6), then for $\varepsilon > 0$ there exists
$\delta > 0$ such that each δ-equivalence $f : M' \to M$ is ε-close
to a homeomorphism $h : M' \to M$, i.e., $\forall x \in M'$
$d(f(x),h(x)) < \varepsilon$. In other words, the answer of Question 1
is affirmative for $F = pt$ and M being a manifold of sufficiently high dimension. Recently, Quinn [8] proved his
important Thin h-cobordism Theorem. Essentially, this theorem
asserts that if M satisfies some minor local connectivity
conditions, E and E' being ANR's, and $Wh(\pi_1 F \times Z^i) = 0$
for all i, then for $\varepsilon > 0$ sufficiently small the answer
of Question 2 is always affirmative, i.e., f is simple.

Modifying the proof of [1], we shall give an exposition of
a structure set analogue of Chapman-Ferry-Quinn theory when
M is a manifold and F is a certain $K(\pi,1)$-manifold, i.e.,
a partial answer to Question 1 for this special case. By F
being a $K(\pi,1)$-manifold, we mean that F is a connected
manifold (possibly with boundary) with $\pi_1 F = \pi$ and $\pi_i F = 0$
for $i > 1$. We shall consider F satisfying

Condition (*): F *is a* $K(\pi,1)$-*manifold of dimension* k
and every homotopy equivalence of manifolds

$$f: (W^{k+i+j}, \partial W^{k+i+j}) \to (F \times D^i \times T^j, \partial(F \times D^i \times T^j))$$

$(i \geq 0, \ j \geq 0, \ k+i+j \geq 5$ *and* $T^j = j$-*dim torus*)

such that $f|\partial W$ *is a homeomorphism, is homotopic (keeping*
$f|\partial W$ *fixed) to a homeomorphism.*

THEOREM A: *Let* $F \to E \overset{p}{\to} M$ *be a locally trivial fibration*
of manifolds such that F *is a* $K(\pi,1)$-*manifold satisfying*
Condition (*) *and* M *is a Riemannian manifold with metric* d.

Let f : E' → E *be an* ε-*homotopy equivalence with respect to*
p *such that* E' *is a manifold and* p|∂E' *is a homeomorphism.*
If dim E ≥ 5, *then there exists* ε_0 > 0 *depending only on*
M *such that if* ε < ε_0 *then* f *is homotopic (keeping* f|∂W
fixed) to a homeomorphism.

We were informed by F. Quinn that it also follows from an
argument using his "End Theorem" of [8]. Apparently, the main
theorem was proved in T. A. Chapman's recent paper "Approximation Results in Finite Dimensional Manifolds" (submitted to
Trans. AMS). Some applications of this theorem will be given
in [3] and [4].

II. A CONTROLLED ENGULFING LEMMA

Let F be a compact manifold (possibly with boundary) of
dimension k. Let W be a manifold of dimension k + ℓ ≥ 5
and let f : W → F × $S^{\ell-1}$ × R be a proper map such that
f|∂W : ∂W → ∂(F × $S^{\ell-1}$ × R) is a homeomorphism and f is an
ε-equivalence over $S^{\ell-1}$ × [-2,2] with respect to the canonical projection p : F × $S^{\ell-1}$ × R → $S^{\ell-1}$ × R and the standard
metric on $S^{\ell-1}$ × R, i.e., there is a homotopy inverse

$$g : F \times S^{\ell-1} \times [-2,2] \to W$$

and homotopies H from gf to $\text{id}_{(fp)^{-1}(S^{\ell-1} \times [-2,2])}$ and
G from fg to $\text{id}_{F \times S^{\ell-1} \times [-2,2]}$ satisfy

(i) ∀ x' ∈ $(pf)^{-1}(S^{\ell-1} \times [-2,2])$ the path (pf)H(x' × I) is
in an ε-neighborhood of every y ∈ pfH(x' × I), i.e.,
diam pfH(x' × I) < ε,

(ii) ∀ x ∈ F × $S^{\ell-1}$ × [-2,2], diam pG(x × I) < ε. Moreover,
g = f^{-1} and H, G are the identity maps when these maps are
restricted to the boundaries.

Let us first consider a collection of chambers $\{U_i\}_{i=1}^m$ of W obtained as follows: let

$$-1 < t_1 < \cdots < t_m < 1$$

be a partition of $(-1,1)$ and let

$$U_i = (pf)^{-1}(S^{\ell-1} \times [t_i, t_{i+1}]) \quad (1 \le i \le m) \quad .$$

Define

$$V_i = (pf)^{-1}(S^{\ell-1} \times \{t_i\}) \quad (1 \le i \le m+1) \quad .$$

The collection $\{U_i\}_{i=1}^m$ is said to be *nice* if $|t_{i+1} - t_i| > 3\varepsilon$.

For $A \subset B \subset C \subset X$ we say that C *deforms into* B *rel* A in X if there is a homotopy

$$H_t : C \to X \qquad (t \in [0,1])$$

such that

   (i) H_0 is the inclusion $C \subset X$,

   (ii) $H_1(C) \subset B$,

   (iii) $H_t(a) = a$ for all $(a,t) \in A \times [0,1]$.

We say that the deformation is ε-*controlled* with respect to a projection $p : X \to Y$ (Y a metric space) if $\forall x \in C$, $pH(X \times I)$ is in an ε-neighborhood of every $y \in pH(X \times I)$.

   <u>LEMMA 2.1.</u> *If* $\{U_i\}_{i=1}^m$ *is a nice collection of chambers in* W, *then*

   (i) *there is a deformation of* $U_i \cup U_{i+1}$ *into* U_i *rel* V_i *in* $U_i \cup U_{i+1} \cup U_{i+2}$;

   (ii) *there is a deformation of* $U_{i+1} \cup U_{i+2}$ *into* U_{i+2} *rel* V_{i+3} *in* $U_i \cup U_{i+1} \cup U_{i+2}$. *Moreover, if* $\varepsilon \ll \varepsilon_1$, *we can make the deformations* ε_1-*controlled over the* $S^{\ell-1}$ *factor.*

Proof.   We shall only construct the deformation of (i).
The deformation of (ii) is obtained in a similar way.

Let

$$a_i = \frac{2}{3} t_i + \frac{1}{3} t_{i+1} \ , \qquad b_i = \frac{1}{3} t_i + \frac{2}{3} t_{i+1}$$

for each $1 \leq i \leq m$.   Next, let

$$H_t : (pf)^{-1} (S^{\ell-1} \times [-2,2]) \rightarrow W$$

be an ε-homotopy with respect to pf from the inclusion map
to the restriction $gf \mid (pf)^{-1}(S^{\ell-1} \times [-2,2])$.   Now, define a
homotopy

$$S_t : F \times S^{\ell-1} \times [-2,2] \rightarrow F \times S^{\ell-1} \times [-2,2] \qquad (t \in [0,1])$$

by

$$S_t(x,u) = (x,(1-t)u + tb_i) \ , \qquad (x,u) \in (F \times S^{\ell-1}) \times [-2,2]$$

Finally, define a homotopy

$$h_t : (pf)^{-1}(S^{\ell-1} \times [b_i,t_{i+2}]) \rightarrow (pf)^{-1}(S^{\ell-1} \times [a_i,t_{i+3}])$$

by

$$h_t(x) = \begin{cases} H_{2t}(x) & (0 \leq t \leq \tfrac{1}{2}) \ , \\[2em] gS_{2t-1} f(x) & (\tfrac{1}{2} \leq t \leq 1) \ . \end{cases}$$

Then h_t is a homotopy from the inclusion to the map $gS_1 f$,
with the following two properties:

(1) $gS_1 f(pf)^{-1}(S^{\ell-1} \times [b_i,t_{i+2}])$
      $\subset (pf)^{-1}(S^{\ell-1} \times [a_i,t_{i+1}])$ and

(2) $h_t \mid (pf)^{-1}(S^{\ell-1} \times \{b_i\})$ is an

      ε-homotopy with respect to pf.

We can extend the restriction

$h_t \mid (pf)^{-1}(S^{\ell-1} \times \{b_i\})$ to a homotopy

$$h_t^1 : (pf)^{-1}(S^{\ell-1} \times [t_i, b_i]) \to (pf)^{-1}(S^{\ell-1} \times [t_i, t_{i+1}])$$

such that

$$h_t^1 = id \quad on \quad (pf)^{-1}(S^{\ell-1} \times [t_i, a_i])$$

and

$$h_t^1 = h_t \quad on \quad (pf)^{-1}(S^{\ell-1} \times \{b_i\}) \quad .$$

Moreover, if $\varepsilon \ll \varepsilon_1$, we can make these deformations ε_1-controlled over the $S^{\ell-1}$ factor. Piecing h_t^1 and h_t together we get the required deformation of $(pf)^{-1}(S^{\ell-1} \times [t_i, t_{i+2}])$ into $(pf)^{-1}(S^{\ell-1} \times [t_i, t_{i+1}])$.

LEMMA 2.2. *There exists an integer* $N = N(\ell)$ *such that for every nice collection* $\{U_i\}_{i=1}^m$ *of chambers in* W *and each integer* k, *such that* $2N + k \leq m$, *there is an isotopy* $h_t : W \to W$, *supported by* $U_1 \cup \cdots \cup U_{2N+k}$, *such that* $h_0 = id$ *and* $h_1(U_1 \cup \cdots \cup U_N) \supset U_1 \cup \cdots \cup U_{N+k}$. *Moreover, if* $\varepsilon \ll \varepsilon_1$ *we can make* h_t ε_1*-controlled over the* $S^{\ell-1}$ *factor.*

Proof. By Lemma 2.1 and engulfing, which can always be chosen carefully enough so that it is ε_1-controlled in the $S^{\ell-1}$ factor if the given number $\varepsilon \ll \varepsilon_1$. This argument is essentially carried out in [8; §5-§6].

LEMMA 2.3. *There exists an integer* $N = N(\ell)$ *such that for every nice collection* $\{U_i\}_{i=1}^{8N}$ *of chambers in* W *there exists a homeomorphism* h $: W \to W$ *with support in* $U_1 \cup \cdots \cup U_{8N}$ *such that*

(i) $U_1 \cup \cdots \cup U_{3N} \subset h(U_1 \cup \cdots \cup U_N) \subset U_1 \cup \cdots \cup U_{4N}$,

(ii) $U_1 \cup \cdots \cup U_{5N} \subset h^2(U_1 \cup \cdots \cup U_N) \subset U_1 \cup \cdots \cup U_{6N}$,

(iii) $U_1 \cup \cdots \cup U_{7N} \subset h^3(U_1 \cup \cdots \cup U_N) \subset U_1 \cup \cdots \cup U_{8N}$.

Moreover, if $\varepsilon \ll \varepsilon_1$ *we can make* h ε_1-*controlled over the* $S^{\ell-1}$ *factor.*

Proof.   By Lemma 2.2, there exists an integer $N = N(\ell)$ such that there are homeomorphisms

$$h_0, h_1, h_2 : W \to W$$

satisfying the following three conditions:

(1) h_0 is supported by $U_1 \cup \cdots \cup U_{4N}$

and $h_0(U_1 \cup \cdots \cup U_N) \supset U_1 \cup \cdots \cup U_{3N}$

(2) h_1 is supported by $U_{2N+1} \cup \cdots \cup U_{6N}$

and $h_1(U_{2N+1} \cup \cdots \cup U_{3N}) \supset U_{2N+1} \cup \cdots \cup U_{5N}$,

(3) h_2 is supported by $U_{4N+1} \cup \cdots \cup U_{8N}$

and $h_2(U_{4N+1} \cup \cdots \cup U_{5N}) \supset U_{4N+1} \cup \cdots \cup U_{7N}$.

Define the homeomorphism h by $h = h_0 h_1 h_2$. Then

$$h(U_1 \cup \cdots \cup U_N) = h_0(U_1 \cup \cdots \cup U_N) ,$$
$$h^2(U_1 \cup \cdots \cup U_N) = h_1 h_0(U_1 \cup \cdots \cup U_N) , \quad \text{and}$$
$$h^3(U_1 \cup \cdots \cup U_N) = h_2 h_1 h_0(U_1 \cup \cdots \cup U_N) .$$

The properties (1)-(3) of the lemma now follow.

Define $Y = h(U_1 \cup \cdots \cup U_N) - \text{Int}(U_1 \cup \cdots \cup U_N)$ and let $U = \bigcup_{-\infty}^{+\infty} h^i(Y)$, where h is the homeomorphism of Lemma 2.3.

LEMMA 2.4.   U *is an open subset of* W.

<u>Proof</u>. Since $h(V_{N+1}) \subset U_{3N+1} \cup \cdots \cup U_{4N}$ and since the set $U_{2N+1} \cup \cdots \cup U_{5N}$ lies in $Y \cup h(Y)$ it follows that the set $Int(U_{2N+1} \cup \cdots \cup U_{5N})$ is contained in $Y \cup h(Y)$ and contains $h(V_{N+1})$. From this, we easily see that U is open in W.

LEMMA 2.5. *Let* Y *be as defined above. Then*

(1) *there is a deformation of* $Y \cup h(Y)$ *into* Y *rel* V_{N+1} *in* $Y \cup h(Y) \cup h^2(Y)$,

(2) *there is a deformation of* $h(Y) \cup h^2(Y)$ *into* $h^2(Y)$ *rel* $h^3(V_{N+1})$ *in* $Y \cup h(Y) \cup h^2(Y)$.

Moreover, for $\varepsilon \ll \varepsilon_1$, *the deformation can be* ε_1-*controlled in the* $S^{\ell-1}$-*factor.*

<u>Proof</u>. First observe that

(1') $Y \cup h(Y) \subset U_{N+1} \cup \cdots \cup U_{6N}$

(2') $Y \cup h(Y) \cup h^2(Y) \supset U_{N+1} \cup \cdots \cup U_{7N}$

(3') $h^2(Y) \supset U_{6N+1} \cup \cdots \cup U_{7N}$

(4') $(h(Y) \cup h^2(Y)) \cap (U_1 \cup \cdots \cup U_{7N}) \subset U_{3N+1} \cup \cdots \cup U_{6N}$.

By Lemma 2.1, there is a deformation of $U_{N+1} \cup \cdots \cup U_{6N}$ into U_{N+1} rel V_{N+1} in $U_{N+1} \cup \cdots \cup U_{6N+1}$, which can be ε_1-controlled in the $S^{\ell-1}$ factor provided $\varepsilon \ll \varepsilon_1$. It follows from (1') and (2') above that the restriction of such a deformation to $Y \cup h(Y)$ provides (1) of the lemma. Similarly, (2'), (3') and (4') help us establish part (2) of the lemma.

It is easy to check that U has two ends. We shall denote them by $\{+\infty\}$ and $\{-\infty\}$. It follows from Lemma 2.5 that we can find neighborhoods $N(+\infty)$ and $N(-\infty)$ of $\{+\infty\}$ and $\{-\infty\}$, respectively, and satisfy the following conditions:

(1) $\overline{N(+\infty)} \cap \overline{N(-\infty)} = \phi$,

(2)   there exist deformations $h_t^{\pm} : N(\pm\infty) \to U$ such that $h_0^{\pm}$ are the inclusions and $h_1^{\pm}$ are homeomorphisms,

(3)   there exist smaller neighborhoods $N(\pm\infty) \supset N_1(\pm\infty)$ $\supset N_2(\pm\infty)$ such that $h_t^{\pm} \mid N_2(\pm\infty) = \mathrm{id}$ and

$$U - N_1(\mp\infty) \subset h_t^{\pm}(N_1(\pm\infty)) \subset U - N_2(\mp\infty),$$

(4)   the deformations $h_t^{\pm} \mid N_1(\pm\infty)$ are ε_1-controlled with respect to the composition of

$$f : U \to S^{\ell-1} \times \mathbb{R} \quad \text{and the canonical projection}$$

$$S^{\ell-1} \times \mathbb{R} \to S^{\ell-1} \quad \text{if} \quad \varepsilon \ll \varepsilon_1.$$

Therefore, $\{\pm\infty\}$ are tame ends of U and we have a gluing W of U [9]. Using the inverse map $g : F \times S^{\ell-1} \times [-2,2] \to W$ and the ε-homotopy of fg to the identity, we can appropriately glue $F \times S^{\ell-1} \times (-a,a)$ for some $0 < a < 2$, to form $F \times S^{\ell-1} \times S^1$ so that we have a homotopy equivalence

$$\hat{f} : \hat{W} \to F \times S^{\ell-1} \times S$$

with $\hat{f} \mid \partial W$ a homeomorphism. Moroeover, we may assume that $\hat{f}$ is an ε_1-homotopy equivalence with respect to the projection $F \times S^{\ell-1} \times S^1 \to S^{\ell-1}$ if $\varepsilon \ll \varepsilon_1$. We define $\hat{f}$ using the gluing trick of [9] on $U - N_1(\pm\infty)$ and the deformations $h_t^{\pm} \mid N_1(\pm\infty)$ (cf. [8, §5]). So we may assume that $\hat{f}$ agrees with f on an open strip of W as defined in [2, §6]. So, there exists a small positive number b $0 < b \ll a < 2$ and an embedding

$$j : f^{-1}(F \times S^{\ell-1} \times (-b,b)) \longrightarrow \hat{W}$$

such that the following diagram commutes

$$f^{-1}(F \times S^{\ell-1} \times (-b,b)) \xrightarrow{\quad f \quad} F \times S^{\ell-1} \times (-a,a)$$

$$\downarrow j \qquad\qquad\qquad\qquad\qquad \downarrow q$$

$$\hat{W} \xrightarrow{\qquad\qquad \hat{f} \qquad\qquad} F \times S^{\ell-1} \times S^1$$

where q is the natural quotient map of the gluing.

THEOREM 2.6. *Let* $f : W^{k+\ell+m} \to F \times D^\ell \times R^m$ $(k+\ell+m) \leq 5)$ *be
a proper map such that* $f \mid \partial W$ *is a homeomorphism and* f *is
an* ε-*homotopy equivalence over* $[-2,2]^m$ *with respect to the
canonical projection* $p : F \times D^\ell \times R^m \to R^m$. *For* ε *sufficiently small, there exist a homotopy equivalence*

$$g : T^{k+\ell+m} \to F \times D^\ell \times T^m$$

and an embedding

$$j : f^{-1}(F \times D^\ell \times (\tfrac{1}{4},\tfrac{3}{4})^m) \to T^{k+\ell+m}$$

satisfying the following conditions:

(1) $g \mid \partial T$ *is a homeomorphism,*

(2) *the following diagram is commutative*

$$f^{-1}(F \times D^\ell \times (\tfrac{1}{4},\tfrac{3}{4})^m) \xrightarrow{\quad f \quad} F \times D^\ell \times (-1,1)^m$$

$$\downarrow j \qquad\qquad\qquad\qquad\qquad \downarrow q$$

$$T^{k+\ell+m} \xrightarrow{\qquad g \qquad} F \times D^\ell \times T^m$$

where q *is the quotient map of the natural gluing map.*

Proof. We shall prove the theorem by an induction on m.
For $m = 1$, the construction before the statement of the
theorem provides the proof. Assume that the theorem is valid
for $m - 1$. Consider the embedding

$$i : S^{m-1} \times R \to R^m$$

such that $i(D^{m-1} \times R) = (\tfrac{1}{8},\tfrac{7}{8})^m$ where D^{m-1} is the southern hemisphere of S^{m-1}. (We shall identify $S^{m-1} \times R$ with $i(S^{m-1} \times R)$.) For $\varepsilon \ll \varepsilon_1$, it follows from the same construction that we have a homotopy equivalence

$$g_1 : \hat{W}_1 \to F \times D^\ell \times S^{m-1} \times S^1$$

and an embedding

$$j_1 : f^{-1}(F \times D^\ell \times S^{m-1} \times (-1,1)) \to W ,$$

satisfying the following conditions

(1) $g_1 \mid \partial \hat{W}_1$ is a homeomorphism and g_1 is an ε_1-equivalence with respect to the canonical projection

$$p_1 : F \times D^\ell \times S^{m-1} \times S^1 \to S^{m-1} ,$$

(2)   the following diagram is commutative

$$
\begin{array}{ccc}
f^{-1}(F \times D^\ell \times S^{m-1} \times (-1,1)) & \xrightarrow{\ f\ } & F \times D^\ell \times S^{m-1} \times (-1,1) \\
\downarrow{\scriptstyle j_1} & & \downarrow{\scriptstyle q_1} \\
\hat{W}_1 & \xrightarrow{\ g_1\ } & F \times D^\ell \times S^{m-1} \times S^1
\end{array}
$$

where q_1 is the quotient map of the canonical gluing.

Consider the proper map

$$g_1 : W_1 = \hat{W}_1 - g_1(F \times D^\ell \times p \times S^1) \to F \times D^\ell \times (S^{m-1} - p) \times S^1$$

where p is the north pole of S^m. If we identify $S^{m-1} - p$ with R^{m-1}, we may apply the induction hypothesis after replacing ε by ε_1. So, if we have ε sufficiently small which also makes ε_1 small, our theorem follows from the induction.

III. THE HANDLE THEOREM

Let us now assume that F is a $K(\pi,1)$-manifold satisfying Condition (*) of the Introduction.

LEMMA 3.1. (The Handle Lemma) *Let* V *be a manifold of dimension* ≥ 5, *and let* $f : V \to F \times D^k \times R^m$ *be a proper map such that* $\partial V = f^{-1}(\partial(F \times D^k \times R^m))$ *and* f *is a homeomorphism over* $\partial F \times (-1,1] \times D^k \times R^m \cup F \times (D^k - \frac{1}{2} D^k) \times R^m$ *where* $\partial F \times (-1,1]$ *denotes a collar of* ∂F *in* F. *Let* $q : F \times D^k \times R^m \to D^k \times R^m$ *be the canonical projection. Then for every* $\varepsilon > 0$ *there exists* $\delta > 0$ *such that if* f *is a* δ-*homotopy equivalence over* $D^k \times 3D^m$ *then*

(i) *there exists an* ε-*homotopy equivalence* $\hat{f} : F \times D^k \times R^m \to F \times D^k \times R^m$ *such that* $\hat{f} = \mathrm{id}$ *over*

$$\partial F \times (-1,1] \times D^k \times R^m \cup F \times ((D^k - \frac{5}{8} \mathring{D}^k)$$

$$\times R^m \cup D^k \times (R^m - 4\mathring{D}^m)) \ ,$$

(ii) *there exists a homeomorphism* $\phi : f^{-1}(U) \to \hat{f}^{-1}(U)$ *such that* $\hat{f}\phi = f \mid f^{-1}(U)$ *where*

$$U = \partial F \times (-1,1] \times D^k \times R^m \cup F \times ((D^k - \frac{5}{8} \mathring{D}^k)$$

$$\times R^m \times D^k \times 2D^m)) \quad .$$

Proof. Let $e^m : R^m \to T^m = S^1 \times \cdots \times S^1$ be the exponential map $e^m = e \cdots e$ where $e : R \to S^1$ is the map $e(t) = e^{\pi i t/4}$, $(t \in R)$. We consider the following modified diagram of [6] [7]

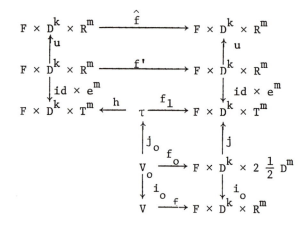

(1) Constructions of V_o and f_o. Let $V_o = f^{-1}(F \times D^k \times 2\frac{1}{2}\overset{\circ}{D}{}^m)$ and $f_o = f \mid V_o$.

(2) Constructions of τ, j_o, f_1 and j. Applying Theorem 2.6 to f, we have a manifold τ and a homotopy equivalence $\tau \to F \times D^k \times T^m$ where τ and $F \times D^k \times T^m$ are gotten from gluings. After appropriate changing of the scales, we may assume that j_o is the map j, f_1 is the map g and j is the restriction of the quotient map q of Theorem 2.6 (such that j is also an embedding). If f is a δ-equivalence over $D^k \times 3D^m$, then we may assume that f_1 is a δ_1-equivalence over $D^k \times T^m$.

(3) Construction of h. It follows from Condition (*) that we can construct a homeomorphism $h : \tau \to F \times D^k \times T^m$ which agrees with f_1 over $\partial F \times (-1,1] \times D^k \times T^m \cup F \times (D^k - \frac{5}{6}\overset{\circ}{D}{}^k) \times T^m$ and which is homotopic to f_1.

(4) Construction of u. Define $u : R^k \times R^m \to 4\overset{\circ}{D}{}^k \times 4\overset{\circ}{D}{}^m$ to be the radial projection which is fixed on $2D^k \times 2D^m$. Then, define $u : F \times D^k \times R^m \to F \times D^k \times R^m$ to be the restriction $u = id_F \times (u' \mid D^k \times R^m)$. Clearly, u is an open embedding.

(5) Construction of $\hat{f}'$. We define f_1' to be the lifting of $f_1 h^{-1}$ which is the identity on $\partial F \times (-1,1] \times T^m \cup F \times (D^k - \frac{5}{6} \mathring{D}^k) \times T^m$. Since $f_2 h^{-1}$ is homotopic to the identity, it follows that f_1' is bounded, i.e., $\| q \hat{f}_1'(x) - q(x) \|$ for $x \in F \times D^k \times R^m$ is bounded where $\| \ \|$ is the standard metric on R^{k+m}. If $\delta_1 > 0$ is small, then we can make $\hat{f}_1'$ an ε_0-equivalence. (Note that δ_1 is small if δ is sufficiently small.) Using the boundedness of f_0, we can find such $\varepsilon_1 > \varepsilon_2 > \varepsilon_3 > 0$ that

$$\hat{f}_1' u^{-1} (F \times D^k \times \partial(4 - \varepsilon_2)D^m)$$
$$\subset u^{-1}(F \times D^k \times ((4 - \varepsilon_3)D^m - (4 - \varepsilon_1)D^m)) \ .$$

Squeezing $(4 - \varepsilon_3)D^m - (4 - \varepsilon_1)D^m$ to $\partial((4 - \varepsilon_2)D^m)$ we can modify $\hat{f}_1'$ to obtain an $2\varepsilon_0$-homotopy equivalence $\hat{f}_2'$ such that

$$\hat{f}_2' u^{-1}(F \times D^k \times \partial(4 - \varepsilon_2)D^m) = u^{-1}(F \times D^k \times \partial(4 - \varepsilon_2)D$$

and $\hat{f}_2' u^{-1}$ agrees with $\hat{f}_1' u^{-1}$ on $u^{-1}(F \times D^k \times (4 - 2\varepsilon_0)D^m)$. If we have chosen ε_0 sufficiently smaller than ε, then $\hat{f}_2'$ is ε-homotopic to a map $\hat{f}' : F \times D^k \times R^m \to F \times D^k \times R^m$ such that $\hat{f}' u^{-1} \mid F \times D^k \times (4D^m - (4 - \varepsilon_2)D^m) = $ id and $\hat{f}' u^{-1}$ agrees with $\hat{f}_1' u^{-1}$ on $u^{-1}(F \times D^k \times (4 - 3\varepsilon_0)D^m)$. (Note that we usually choose ε_1 much smaller than ε_0.)

(6) Construction of $\hat{f}$. $F \times D^k \times R^m \to F \times D^k \times R^m$ as follows:

$$\hat{f}(x) = \begin{cases} u\hat{f}'u^{-1}(x) & \text{for } x \in u(F \times D^k \times R^m) \\[2em] x & \text{otherwise .} \end{cases}$$

(7) Construction of ϕ. It is easy to check that we have a commutative diagram

$$\hat{f}^{-1}(F \times D^k \times 2D^m) \xrightarrow{\quad \hat{f} \quad} F \times D^k \times 2D^m$$

$$\downarrow h^{-1}(\mathrm{id} \times e^m) \qquad\qquad\qquad \downarrow \mathrm{id} \times e^m$$

$$(f_0^{-1}(\mathrm{id} \times e^m))(F \times D^k \times 2D^m) \xrightarrow{\quad f_0 \quad} (\mathrm{id} \times e^m)(F \times D^k \times 2D^m)$$

$$\downarrow i_0 \qquad\qquad\qquad\qquad\qquad \downarrow i_1$$

$$f^{-1}(F \times D^k \times 2D^m) \xrightarrow{\quad f \quad} F \times D^k \times 2D^m \quad .$$

The vertical arrows are homeomorphisms and by composing the inverses of the two on the left we get a homeomorphism $\psi : f^{-1}(F \times D^k \times 2D^m) \to \hat{f}^{-1}(F \times D^k \times 2D^m)$, for which $\hat{f}\psi = f \mid f^{-1}(F \times D^k \times 2D^m)$. We now extend ψ to a homeomorphism $\phi : f^{-1}(U) \to \hat{f}^{-1}(U)$ by setting $\phi = f$ on $\partial F \times (-1,1] \times D^k \times R^m \cup F \times (D^k - \frac{5}{8}\mathring{D}^k) \times R^m$.

THEOREM 3.2. (The Handle Theorem) *Let* V *be a manifold of dimension* ≥ 5 *and let* $f : V \to F \times D^k \times R^m$ *be a proper map such that* $\partial V = f^{-1}(\partial(F \times D^k \times R^m))$ *and* f *is a homeomorphism on* $\partial F \times (-1,1] \times D^k \times R^m \cup F \times (\mathring{D}^k - \frac{1}{2}D^k) \times R^m$, *where* $\partial F \times (-1,1]$ *is a collar of* ∂F *in* F. *Then, for every* $\varepsilon > 0$ *there exists* $\delta > 0$ *such that if* f *is a δ-equivalence over* $D^k \times 3D^m$ *(with respect to the natural projection* $F \times D^k \times R^m \to D^k \times R^m$*) then there is proper map* $f : V \to F \times D^k \times R^m$ *such that*

(1) $\tilde{f}$ *is an ε-equivalence over* $D^k \times \frac{5}{2}D^m$,

(2) $\tilde{f} = f$ *over*

$$\partial F \times (-1,1] \times D^k \times R^m \cup F \times ((D^k - \frac{2}{3}\mathring{D}^k)$$

$$\times R^m \times D^k \times (R^m - 2\mathring{D}^m) \, ,$$

(3) $\tilde{f}$ *is a homeomorphism over* $F \times D^k \times D^m$.

Proof. For $m = 0$, it follows from Condition $(*)$. Therefore assume $m \geq 1$. For any $\delta_1 > 0$ we can choose $\delta > 0$
small enough so that if the map f is a δ-homotopy equivalence over $D^k \times 3D^m \subset D^k \times R^m$, then there exists
$\hat{f} : F \times D^k \times R^m \to F \times D^k \times R^m$ and $\phi : f^{-1}(U) \to \hat{f}^{-1}(U)$ as
described in Lemma 3.1 if ε is replaced by δ_1. Consider
the restriction

$$\hat{f}| : F \times D^k \times R^m - f^{-1}(F \times D^k \times 0) \to F \times D^k \times (R^m - 0) .$$

For any compact subset Y of $D^k \times (R^m - 0)$, δ_1 can be
chosen small enough so that this restriction $\hat{f}|$ is a δ_1-
equivalence over Y. Since $f = $ id over a neighborhood of ∞
we can identify S^m with $R^m \cup \{\infty\}$ and extend $\hat{f}|$ to
$\hat{f}_1 : V_1 \to F \times D^k \times (S^m - 0)$, where V_1 is a manifold. This
extension can be carried out so that $\hat{f}_1$ is a homeomorphism
over $\partial F \times (-1,1] \times D^k \times R^m \cup F \times (D^k - \frac{5}{6} D^k) \times (S^m - 0)$.
Also $\hat{f}_1$ will be a δ_1-equivalence over any conveniently chosen
compact subset in $D^k \times (S^m - 0)$ (cf. [10]). By Lemma 3.1,
there exists a δ_2-equivalence $f_2 : F \times D^k \times (S^m - 0) \to F \times D^k$
$\times (S^m - 0)$ such that $\hat{f}_2 = $ id on

$$\partial F \times (-1,1] \times D^k \times R^m \cup F \times \{[(D^k - \frac{6}{7} \mathring{D}^k)$$
$$\times (S^m - 0)] \cup (D^k \times (\mathring{D}^m - 0)\}$$

and there is a homeomorphism $\phi_1 : \hat{f}_1^{-1}(U_1) \to \hat{f}_2^{-1}(U_2)$ such
that $\hat{f}_2 \phi_1 = \hat{f}_1 | \hat{f}_1^{-1}(U_1)$ where

$$U_1 = \partial F \times (-1,1] \times D^k \times R^m \cup F \times \{[(D^k - \frac{6}{7} \mathring{D}^k)$$
$$\times (S^m - 0)] \cup D^k \times (S^m - \frac{3}{2} \mathring{D}^m)\}$$

and $\hat{f}_2 \phi_1 \phi = f$ over

$$F \times \{[(D^k - \frac{6}{7} \mathring{D}^k) \times (R^m - 0)] \cup (2D^m - \frac{3}{2} \mathring{D}^m)\} .$$

Extend $\hat{f}_2$ to $\hat{f}_2 : F \times D^k \times S^m \to F \times D^k \times S^m$ by defining
$\hat{f}_2 | F \times D^k \times 0 = $ id. Then f_2 is still a δ_2-equivalence.

Now, consider the open subset $G = F \times ((\frac{7}{8} \mathring{D}^k \times 2\mathring{D}^m)$
$- (\frac{6}{7} D^k \times \frac{3}{2} D^m))$ which is homeomorphic to $F \times S^{m+k-1} \times R$.
If we choose ε small enough then there is bicollared sub-
manifold S of $f^{-1}(G)$, which is homeomorphic to $F \times S^{m+k-1}$.
In fact, we can find a bicollared submanifold S' of
$\tilde{f}_2^{-1}(G) \subset \tilde{f}_2^{-1}(F \times D^k \times 2D^m)$ homeomorphic to $F \times S^{m+k-1}$ and
we may set $S = (\phi_1 \phi)^{-1}(S')$. It follows from Condition (*)
that S bounds a manifold (homeomorphic to) $F \times D^{m+k}$ in
$f^{-1}(F \times D^k \times 2\mathring{D}^m)$ which contains $f^{-1}(F \times \frac{6}{7} D^k \times \frac{3}{2} D^m)$. Let
B' be the manifold in $\tilde{f}_2^{-1}(F \times D^k \times 2D^m)$ bounded by S'. We
have a homeomorphism $h : F \times D^{m+k} \to B'$ so that $h \mid S = \phi_1 \phi$.
We define

$$\tilde{f} : V \to F \times D^k \times R^m$$

by

$$\tilde{f}(x) = \begin{cases} f(x) & \text{for } x \in V - F \times D^{m+k} , \\ \\ \tilde{f}_2 h(x) & \text{for } x \in F \times D^{m+k} . \end{cases}$$

So $\tilde{f} = f$ over

$$\partial F \times (-1,1] \times D^k \times R^m \cup F \times \{((D^k - \frac{7}{8} \mathring{D}^k) \times R^m)$$
$$\cup (D^k \times (R^m - 2\mathring{D}^m))\}$$

and $\tilde{f}$ is a homeomorphism over $F \times D^k \times D^m$ rel $\partial F \times (-1,1]$
$\times D^k \times D^m \cup F \times \partial D^k \times (-a,a] \times D^m$ where $\partial D^k \times (-a,a]$ is a
collar neighborhood of ∂D^k in D^k. We can easily make $\tilde{f}$
an ε-homotopy equivalence. Finally, we replace $\frac{7}{8}$ by $\frac{2}{3}$ and
the theorem follows.

IV. PROOF OF THEOREM A

In fact, we have a stronger theorem. Let us first introduce
the following notation. For a space M with metric d, a
subset C of M and a given $\alpha > 0$, we set

$$C^{\alpha} = \{x \mid d(x,C) < \alpha\} \ .$$

THEOREM 4.1. *Let* $F \to E \to M$ *be a locally trivial fibration of manifolds such that* M *is a Riemann manifold with metric* d *and* F *is a* $K(\pi,1)$-*manifold satisfying Condition* (*).

Then given compact subsets $C_1 \subset C_2 \subset M$ *and small numbers* $\alpha > 0, \ \varepsilon > 0,$ *there is* $\delta = \delta(\varepsilon,C_1,C_2)$ *such that every* δ-*equivalence* $f : E' \to E$ *with*

$$f \mid (pf)^{-1}(C_1^{2\alpha}) = (pf)^{-1}(C_1^{2\alpha}) \to p^{-1}(C_1^{2\alpha})$$

a homeomorphism is $\frac{1}{2}\varepsilon$-*homotopic to an* ε-*equivalence* $f' : E' \to E$ *so that*

    (i) $f' \mid (pf')^{-1}(C_2^{\alpha}) : (pf')^{-1}(C_2^{\alpha}) \to p^{-1}(C_2^{\alpha})$

    is a homeomorphism.

    (ii) $f' = f$ *over* $(pf)^{-1}(C_1^{\alpha})$.

Proof. Consider compact submanifolds $\overline{C}_1^{\alpha} \subset M_1 \subset C_1^{2\alpha}$ and $M_1 \cup \overline{C}_2^{\alpha} \subset M_2 \subset C_2^{2\alpha}$ such that M_2 is gotten from M_1 by attaching a finite number of handles. We may repeatedly apply Theorem 3.2 as follows. Inductively, we may assume that we have modified f over some handles and we wish to extend our modification over one more handle. Identify this handle with $F \times D^k \times D^m$ (dim M = k + m), and enlarge it slightly to an open handle which is identified as $F \times D^k \times R^m$. Apply Theorem 3.2 to this open handle. Of course, the number δ of Theorem 3.2 gets smaller and smaller when we move on from one handle to the next. Since M_2 has only finite number of handles over M_1, we eventually produce a number $\delta = \delta(\varepsilon,C_1,C_2)$ and $f' : E' \to E$ satisfying the theorem. This proves Theorem 4.1.

Clearly, Theorem 4.1 implies Theorem A.

ACKNOWLEDGMENTS

Partial support by the National Science Foundation under grant number MCS-7701124 is acknowledged by the first-named author. Partial support by the National Science Foundation under grant number GP 34324X1 is acknowledged by the second-named author.

REFERENCES

1. Chapman, T. A. and S. Ferry, "Approximating Homotopy Equivalences by Homeomorphisms." <u>Amer. J. Math</u>. 101 (1979), 583-607.

2. Farrell, F. T. and W. C. Hsiang, "Manifolds with $\pi_1 = G \times_\alpha T$." <u>Amer. J. Math</u>. 95 (1973), 813-848.

3. ------, "Topological Characterization of Flat and Almost Flat Riemannian Manifolds $M^n(n \neq 3,4)$." (To appear.)

4. ------, "On Novikov's Conjecture for Non-positively Curved Manifolds II." (To appear.)

5. Ferry, S., "Homotopy ε-maps to Homeomorphisms." <u>Amer. J. Math</u>. 101 (1979), 567-582.

6. Kirby, R., "Stable Homeomorphisms and Annulus Conjecture." <u>Ann. of Math</u>. 89 (1969), 575-582.

7. Kirby, R. and L. C. Siebenmann, "Foundational Essays on Topological Manifolds, Smoothings, and Triangulations." <u>Ann. of Math. Studies</u> 88, Princeton Univ. Press (1977).

8. Quinn, F., "Ends of Maps." <u>Ann. of Math</u>. 110 (1979), 275-331.

9. Siebenmann, L. C., "A Total Whitehead Obstruction." <u>Comment. Math. Helv</u>. 45 (1970), 1-48.

10. ------, "Approximating Cellular Maps by Homeomorphisms." <u>Topology</u> 11 (1973), 271-294.

Heegaard Splittings of Homology Three-Spheres and Homotopy Three-Spheres

D. R. McMillan, Jr.

1. INTRODUCTION

A regular neighborhood in $\mathbb{R}^3$ of a finite, connected graph of Euler characteristic $1-n$ is called a *handlebody of genus* n. Each closed orientable 3-manifold M^3 is the union of two such "Heegaard handlebodies" K_1, K_2 with $\partial K_1 = K_1 \cap K_2 = \partial K_2$. Let $D_1, \cdots, D_n$ be a disjoint collection of disks properly embedded in K_2 and cutting K_2 into a 3-cell. If for some embedding $\phi : K_1 \to S^3$, each $\phi(\partial D_i)$ is homotopic to a constant in $S^3 - \phi(\mathring{K}_1)$ then M^3 is topologically S^3. Several authors have considered approximations to such an ideal embedding ϕ. W. Haken has shown that M^3 is a homotopy 3-sphere if and only if ϕ can be chosen so that the $\phi(\partial D_i)$'s bound disjoint orientable surfaces in $S^3 - \phi(\mathring{K}_1)$. (See [5], [6], and Bing's [1]. This result can also be deduced from Moise's [9].) D. Neumann has shown that M^3 is a homology 3-sphere if and only if there is such a ϕ with each $\phi(\partial D_i)$ homologous to zero in $S^3 - \phi(\mathring{K}_1)$. (See [10].)

We give a unified proof of these results of Haken and Neumann. (We consider just the "only if" claims, however.) In addition, we extend Neumann's result to show that if M^3 is a homology 3-sphere and p is a positive integer, then there is an embedding $\phi : K_1 \to S^3$ such that each $\phi(\partial D_i)$ belongs to the pth derived group of $\pi_1(S^3 - \phi(\mathring{K}_1))$. (Taking $p = 1$ gives Neumann's result.) This last may be a little surprising but is, on the whole, a discouraging outcome: Homology 3-spheres cannot be distinguished by studying their (p)-*embeddings* (in the terminology of [10]). Further, having a (p)-embedding for each p does not guarantee that M^3 is simply connected. The obvious question remains unanswered: If there is a *single* embedding $\phi : K_1 \rightarrow S^3$ such that each $\phi(\partial D_i)$ belongs to the pth derived group of $\pi_1(S^3 - \phi(\mathring{K}_1))$ for each positive integer p, must M^3 be simply connected? (A similar question can be asked for the lower central series of subgroups.)

We also make an effort toward proving that each homotopy 3-sphere M^3 embeds in S^4. We show that the quotient of M^3 by some cell-like, essentially 1-dimensional set embeds in S^4, and that M^3 minus some compact, 0-dimensional set has a bicollared embedding in $\mathbb{R}^4$. It seems reasonable to conjecture that each open, acyclic 3-manifold embeds in $\mathbb{R}^4$. (The usual such examples are monotone unions of handlebodies, and one can construct bicollared embeddings of these in $\mathbb{R}^4$ without use of the "acyclic" hypothesis.)

We work throughout in the piecewise-linear (PL) category. Our notation is as follows: $\mathbb{R}^n$ for Euclidean n-space; S^n for the unit sphere in $\mathbb{R}^{n+1}$; H_* for integral singular homology; χ for Euler characteristic; and $\approx$ for homeomorphism of spaces. $\mathring{M}^n$ denotes the *interior* of an n-manifold M^n; and $\overset{\circ}{\subset}$ means "contained in the interior of." A *cell-like set* is by definition compact, metrizable, and admits no essential mapping into a finite polyhedron. Finally, the *derived series* of a group G is defined inductively by $G^{(0)} = G$ and $G^{(p)} = [G^{(p-1)}, G^{(p-1)}]$ for $p \geq 1$, where [A,B] is the subgroup of G generated by all *commutators* $[a,b] = a^{-1}b^{-1}ab$ (a $\in$ A, b $\in$ B).

2. EMBEDDINGS OF HEEGAARD HANDLEBODIES IN S^3

__LEMMA:__ *Suppose that the integral homology 3-sphere* M^3 *is the union of handlebodies* K_1, K_2 *with*

$$\partial K_1 = K_1 \cap K_2 = \partial K_2 \quad .$$

Then for some PL *self-homeomorphism* h *of* M^3, h *is isotopic to the identity and the inclusion* $K_1 \to h(\mathring{K}_1)$ *induces zero on integral first homology. If* $\pi_1(M^3)$ *is trivial, then* h *can be chosen so that each of the inclusions* $K_1 \to h(\mathring{K}_1)$ *and* $h(K_2) \to \mathring{K}_2$ *is homotopic to a constant.*

__Proof.__ Since $H_1(M^3) = 0$, it follows from any one of [14; Thm. 3], [5; Sect. 6], or [13; Corollary] that the inclusion of K_1 into some handlebody $V \subset M^3$ induces zero on first homology. (If $\pi_1(M^3) = \{1\}$, each of these sources guarantees that V can be chosen to induce the trivial π_1-homomorphism.) Each of the handlebodies K_2, V is a regular neighborhood of a 1-dimensional polyhedron, say A, B respectively. By general position, we can assume that $A \cap B = \emptyset$. Hence, there are self-homeomorphisms f, g of M^3 (isotopic to identity) which squeeze K_2, V so close to A, B (respectively) that $f(K_2) \cap g(V) = \emptyset$. It follows that

$$V \subset g^{-1}f(\mathring{K}_1) \quad .$$

Thus, $g^{-1}f$ is the desired homeomorphism h if we assume only that M^3 is a homology 3-sphere and require only that the inclusion $K_1 \to h(\mathring{K}_1)$ should induce zero on H_1.

If $\pi_1(M^3) = \{1\}$, the above definition of h will give the inessentiality of the inclusion $K_1 \to h(\mathring{K}_1)$, but perhaps not that of the inclusion $h(K_2) \to \mathring{K}_2$. (This last inclusion does, of course, induce zero on H_1.) To remedy this defect, we take more care in choosing h in this case. Let

$h_1 = g^{-1}f$ be as above, so that $K_1 \to h_1(\mathring{K}_1)$ is homotopic to a constant. The above argument then yields a PL self-homeomorphism h_2 of M^3 (isotopic to identity) for which the inclusion

$$h_1(K_2) \to h_2 h_1(\mathring{K}_2)$$

is homotopic to a constant. Letting $h = h_2^{-1} h_1$, we find that

$$h(K_2) \overset{e}{\subset} h_1(K_2) \subset K_2 \quad , \quad \text{and}$$

$$K_1 \overset{e}{\subset} h_1(K_1) \subset h(K_1) \quad ,$$

where the first of each of these inclusions is homotopic to a constant. This completes the proof.

THEOREM 1: *Suppose that the integral homology 3-sphere* M^3 *is the union of handlebodies* K_1, K_2, *with*

$$\partial K_1 = K_1 \cap K_2 = \partial K_2 \quad .$$

Let $D_1, \cdots, D_n$ *be a disjoint collection of disks cutting* K_2 *into a 3-cell* ($n = 1 - \chi(K_2)$). *Let* p *be a given positive integer. Then for some embedding*

$$\phi : K_1 \to S^3 \quad ,$$

each of the curves $\phi(\partial D_i)$ *represents an element of the* p^{th} *derived group of*

$$\pi_1(S^3 - \phi(\mathring{K}_1)) \quad .$$

Proof. The lemma provides a self-homeomorphism h of M^3 for which the inclusion $K_1 \to h(\mathring{K}_1)$ induces zero on integral first homology. By duality, the inclusion $K_2 \to h^{-1}(\mathring{K}_2)$ also induces zero. Hence, each integral 1-cycle in $\partial K_1 = \partial K_2$ is zero-homologous in the complement of each of $h(K_2)$ and $h^{-1}(K_1)$. Note that $h(K_2) \cap h^{-1}(K_1) = \emptyset$. Since $H_2(M^3) = 0$,

it follows from the Alexander Addition Theorem (page 60 of
[15]) that the inclusion

$$\partial K_2 \to h^{-1}(K_2) - h(K_2)$$

induces zero on first homology. By induction, each of the
inclusions

$$h^{p-i}(K_2) - h^{p+i}(\overset{\circ}{K}_2) \to h^{p-i-1}(K_2) - h^{p+i+1}(\overset{\circ}{K}_2) \quad ,$$

(for $i = 0,1, \cdots, p-1$) induces zero on first homology. We
have thus established the point of this paragraph: Each loop
in $h^p(\partial K_2)$ represents, under inclusion, an element of the
p^{th} derived group $G^{(p)}$ of

$$G = \pi_1(K_2 - h^{2p}(\overset{\circ}{K}_2)) \quad .$$

Since each of the loops ∂D_i clearly represents an element
of the normal subgroup of G generated by loops in $h^p(\partial K_2)$,
the previous paragraph implies that each ∂D_i represents an
element of $G^{(p)}$. But $M^3 - h^{2p}(\overset{\circ}{K}_2)$ is a handlebody, so we
can choose ϕ to be the restriction to K_1 of any embedding
of this handlebody in S^3. The proof is complete.

THEOREM 2: (W. Haken [5]) *Suppose that the homotopy*
3-sphere M^3 *is the union of handlebodies* K_1, K_2 *with*

$$\partial K_1 = K_1 \cap K_2 = \partial K_2 \quad .$$

Let $D_1, \cdots, D_n$ *be a disjoint collection of disks cutting* K_2
into a 3-cell (n = 1 - $\chi(K_2)$). *Then for some embedding*

$$\phi : K_1 \to S^3 \quad ,$$

there are disjoint, orientable surfaces $\Sigma_1, \cdots, \Sigma_n$ *in*
$S^3 - \phi(\overset{\circ}{K}_1)$ *with* $\partial \Sigma_i = \phi(\partial D_i)$.

Proof. (We remark that our conclusion implies that for
each p, each of the curves $\phi(\partial D_i)$ represents an element

of the pth term of the lower central series of

$$\pi_1(S^3 - \phi(\overset{\circ}{K}_1)) \quad . \quad)$$

The lemma (applied to K_2) gives a self-homeomorphism h^{-1} of M^3 for which the inclusion $h(K_2) \to \overset{\circ}{K}_2$ is homotopic to a constant. Now $h(K_2)$ is a regular neighborhood of a con- nected 1-dimensional polyhedron Γ. Since Γ contracts to a point in $\overset{\circ}{K}_2$, we can add disjoint orientable handles in $\overset{\circ}{K}_2 - \Gamma$ to the disks $D_1, \cdots, D_n$ to obtain disjoint, orientable, non-singular surfaces $\Delta_1, \cdots, \Delta_n$ in $K_2 - \Gamma$ with $\partial\Delta_i = \partial D_i$ for each i. (Cf. the cancellation argument used to prove Theorem 2 of [7].) Since

$$K_1^* = K_1 \cup \bigcup_{i=1}^{n} \Delta_i \subset M^3 - \Gamma \approx M^3 - h(K_2) \approx \overset{\circ}{K}_1 \quad ,$$

K_1^* embeds in S^3 *via* ϕ, say. We put $\sum_i = \phi(\Delta_i)$ to finish the proof.

3.   CELL-LIKE SETS IN HOMOTOPY THREE-SPHERES

THEOREM 3: *Let* M^3 *be a homotopy 3-sphere. Then* M^3 *contains a cell-like set* X *such that* X *is a nested inter- section of handlebodies, while* $M^3 - X$ *is contractible and is a monotone union of handlebodies.*

Proof. As is well-known, M^3 is the union of handlebodies K_1, K_2 with $\partial K_1 = K_1 \cap K_2 = \partial K_2$. By the lemma, there is a PL self-homeomorphism h of M^3 for which each of the in- clusions $K_1 \to h(\overset{\circ}{K}_1)$ and $h(K_2) \to \overset{\circ}{K}_2$ is homotopic to a constant. The compact set

$$X = \overset{\infty}{\underset{i=1}{\cap}} h^i(K_2)$$

is cell-like, since for each i the inclusion

$$h^{i+1}(K_2) \to h^i(\mathring{K}_2)$$

of its neighborhoods is homotopic to a constant. Also, $M^3 - X$ is the monotone union of handlebodies $h^i(K_1)$, for $i = 1,2, \cdots$. Since the inclusion

$$h^i(K_1) \to h^{i+1}(\mathring{K}_1)$$

is homotopic to a constant for each i, it follows that $M^3 - X$ is contractible. Thus, X is the desired set.

COROLLARY 1: *Let* M^3 *be a homotopy 3-sphere, and let* $\sum M^3$ *be its suspension. Then for some arc* A *in* $\sum M^3$ *that joins the suspension points,*

$$(\sum M^3) - A \approx \mathbb{R}^4 \quad .$$

Proof. Let $X \subset M^3$ be the cell-like set guaranteed by Theorem 3. By any one of [3], [11], or [2], there is a surjective mapping

$$f : \sum M^3 \to \sum M^3$$

which fixes the suspension points, and whose nondegenerate point-inverses are precisely the copies of X lying at different suspension levels. The arc A is $f(\sum X)$. We note that

$$(\sum M^3) - A \approx (\sum M^3) - (\sum X) \approx (M^3 - X) \times \mathbb{R}^1 \quad .$$

Since $M^3 - X$ is contractible and is a monotone union of handlebodies, this last product is topologically $\mathbb{R}^4$ by [8].

COROLLARY 2: *Let* M^3 *be a homotopy 3-sphere, and let* X *be the cell-like set obtained in Theorem 3. Then* M^3 *mod* X (M^3 *with* X *crushed to a point) embeds in* S^4.

Proof. Let $\sum M^3$ be the suspension of M^3. Then $\sum M^3$ contains the cell-like set $\sum X$, and

$$(\sum M^3) - (\sum X) \approx (M^3 - X) \times \mathbb{R}^1 \approx \mathbb{R}^4 \quad ,$$

as in the proof of Corollary 1. Hence, $\sum M^3 \bmod \sum X$ is topologically S^4 and contains $M^3 \bmod X$, as claimed.

REMARK. Let $Y = M^3 \bmod X$. The embedding $g : Y \to S^4$ constructed in Corollary 2 shows the following evidence of being bicollared: $S^4 - g(Y)$ is 1-ULC; $S^4 - g(Y)$ has two components U_1 and U_2, where $U_1 \approx \mathbb{R}^4 \approx U_2$; and for each $\varepsilon > 0$ and each $i \in \{1,2\}$, g can be ε-approximated by a mapping $g_i : Y \to U_i$. If $g(Y)$ *is* bicollared in S^4, then M^3 has a bicollared embedding in S^4 also.

COROLLARY 3: *Each homotopy 3-sphere* M^3 *contains a compact, zero-dimensional set* C *such that*

$$(M^3 - C) \times [-1,1]$$

embeds as a closed set in $\mathbb{R}^4$.

Proof. We use the notation of Corollary 1. Since A does not contain uncountably many disjoint open intervals, the copy M_t^3 of M^3 at some suspension level t has a zero-dimensional intersection C with A. Thus,

$$M_t^3 - C \subset (\sum M^3) - A \approx \mathbb{R}^4 \quad ,$$

so that $(M_t^3 - C)$ embeds as a closed set in $\mathbb{R}^4$. Further, since M_t^3 has a product (with $[-1,1]$) neighborhood in $\sum M^3$, the above embedding of $M_t^3 - C$ into $(\sum M^3) - A$ extends to an embedding of $(M_t^3 - C) \times [-1,1]$ into $(\sum M^3) - A$.

QUESTIONS. Does each homotopy 3-sphere embed in S^4?
Does the conclusion of Corollary 3 remain valid if M^3 is
assumed only to be a homology 3-sphere? Is there a closed,
orientable 3-manifold M^3 so that M^3 minus a point
("*punctured* M^3") fails to embed in S^4, yet M^3 minus some
closed 0-dimensional set does embed in S^4? Similar questions
can be asked with respect to embedding M^3 minus a cell-like
(or merely acyclic) set.

D. B. A. Epstein [4] has used results of D. Puppe [12] to
show that the punctured lens spaces $L(2p,q)$ do not embed
smoothly in S^4. We note below that this result can be extended to show that the conclusion of Corollary 3 fails for
these lens spaces.

PROPOSITION: *Let* M^3 *be one of the lens spaces* $L(2p,q)$.
Let C *be any compact, zero-dimensional set in* M^3. *Then*
$(M^3 - C) \times [-1,1]$ *fails to embed in* S^4.

Proof. Let C be as above, and let U be a compact, PL
3-manifold neighborhood of C for which ∂U is connected
and the inclusion

$$U \to (\text{Some Heegaard handlebody } V \text{ of } M^3)$$

is homotopic to a constant. It suffices to show that
$N^3 \times [-1,1]$ fails to embed in S^4, where the closed
orientable 3-manifold

$$N^3 = \partial([M^3 - \mathring{U}] \times [-1,1]) \quad .$$

(We identify $(M^3 - \mathring{U}) \times \{1\}$ with $M^3 - \mathring{U}$.)
In [12], Puppe defines the concept of an n-manifold's being
spherelike ("sphärenähnlich"). This is a necessary condition
for the n-manifold to embed in S^{n+1}. An appealing formula-
tion of spherelikeness is given in his Satz 12: A manifold

M^n is spherelike if and only if there is a degree-one mapping $L^n \to M^n$, where L^n is a manifold for which $L^n \times [-1,1]$ embeds in S^{n+1}. Thus, in our application it suffices to find a degree-one mapping $N^3 \to M^3$. (M^3 is known to be non-spherelike.)

This is done by starting with a set of cutting disks $D_1, \cdots, D_k$ for V ($k = 1 - \chi(V)$), adding handles (as in the proof of Theorem 2) to the D_i's to obtain disjoint surfaces in $V - U$ with the same boundaries, then using these surfaces and the Tietze theorem to construct a mapping

$$f : N^3 - \text{Int}(M^3 - \overset{\circ}{V}) \to V$$

that restricts to a homeomorphism of boundaries. When f is patched together with the identity map of $M^3 - \overset{\circ}{V}$, the desired map $N^3 \to M^3$ results. The proof is complete.

We remark that the same proof shows that $L(2p,q)$ minus a cell-like set never embeds in S^4.

ACKNOWLEDGMENT

Partial support of this research by an N.S.F. grant is hereby acknowledged.

REFERENCES

1. R. H. Bing, "Mapping a 3-sphere onto a homotopy 3-sphere," in Topology Seminar Wisconsin, 1965 (R. H. Bing and R. J. Bean, Editors) Ann. of Math. Studies, no. 60, pp. 89-99. Princeton Univ. Press, Princeton, N.J., 1966.

2. J. W. Cannon, "$(E^3/X) \times E^1 = E^4$ (X a cell-like set), an alternative proof," Trans. A.M.S. 240 (1978), 277-285.

3. R. D. Edwards and R. T. Miller, "Cell-like closed 0-dimensional decompositions of $\mathbb{R}^3$ are $\mathbb{R}^4$ factors," Trans. A.M.S. 215 (1976), 191-203.

4. D. B. A. Epstein, "Embedding punctured manifolds," Proc. A.M.S. 16 (1965), 175-176.

5. W. Haken, "On homotopy 3-spheres," Illinois J. Math. 10 (1966), 159-178.

6. ------, "Various aspects of the three-dimensional Poincaré problem," in Topology of Manifolds (J. C. Cantrell and C. H. Edwards, Jr., Editors), pp. 140-152. Markham, Chicago, Ill., 1970.

7. W. Jaco and D. R. McMillan, Jr., "Retracting three-manifolds onto finite graphs," Illinois J. Math. 14 (1970), 150-158.

8. D. R. McMillan, Jr., "Cartesian products of contractible open manifolds," Bull. A.M.S. 67 (1961), 510-514.

9. E. E. Moise, "A monotonic mapping theorem for simply connected 3-manifolds," Illinois J. Math. 12 (1968), 451-474.

10. Dean A. Neumann, "Heegaard splittings of homology 3-spheres," Trans. A.M.S. 180 (1973), 485-495.

11. C. Pixley and W. Eaton, "S^1 cross a UV^∞ decomposition of S^3 yields $S^1 \times S^3$," in Geometric Topology (L. C. Glaser and T. B. Rushing, Editors), Lecture Notes in Mathematics, Vol. 438, pp. 166-194. Springer-Verlag, New York, 1975.

12. D. Puppe, "Homotopiemengen und ihre induzierten Abbildungen. II, Sphärenähnliche Mannigfaltigkeiten," Math. Zeitschr. 69 (1958), 395-417.

13. N. Smythe, "Handlebodies in 3-manifolds," Proc. A.M.S. 26 (1970), 534-538.

14. J. H. C. Whitehead, "A certain open manifold whose group is unity," Quart. J. Math. (2) 6 (1935), 268-279.

15. R. L. Wilder, "Topology of Manifolds," Amer. Math. Soc. Colloq. Publ. vol. 32, Amer. Math. Soc., Providence, Rhode Island, 1963.

Shape Properties of Compacta in Generalized n-Manifolds

S. Armentrout and S. Singh

ABSTRACT

It is known that a generalized n-manifold, $n \geq 3$, may not
contain any proper compact subset which is strongly movable.
We prove the following: If G is a 0-dimensional CE u.s.c.
decomposition of an n-manifold M^n with $n \geq 3$, then M^n/G
contains a movable continuum of dimension ≥ 2. Many other
related results are given; for instance, it is shown that X/A
may be non-movable even when X and A are both movable. An
elementary proof of a version of Theorem (19) of R. H. Bing
[Fund. Math. 36 (1949), 303-318] is provided, see Theorem
(3.2.5); and some applications of this result are also given.
It is observed that every generalized cactoid is pointed
movable. Many other related matters are discussed.

1. INTRODUCTION

By an AR (ANR) we mean a compact metric absolute (neighborhood) retract. A finite dimensional ANR X is a *generalized
n-manifold* if for each x in X, the homology groups
$H_*(X, X - \{x\}; Z)$ and $H_*(E^n, E^n - \{0\}; Z)$ are isomorphic
where Z denotes the group of integers under addition and E^n
denotes the n-dimensional Euclidean space. All our n-manifolds will be compact and without boundary. A finite dimensional image of an n-manifold under a cell-like map is a
generalized n-manifold $[WI_2]$. F. Quinn [Q] has announced that
every generalized n-manifold X, for $n \geq 5$, has a resolu-
tion (i.e., there exists a cell-like map from an n-manifold
onto X). J. W. Cannon has identified an important property

known as DDP ("Disjoint Disks Property") cf. $[C_1]$ and R. D.
Edwards [E] has used this property to prove the following
theorem: If f : M → X is a cell-like map from an n-manifold,
n ≥ 5, onto a generalized n-manifold X satisfying DDP,
then f can be (arbitrarily close) approximated by homeomorphisms, i.e., the generalized n-manifold X is an honest
n-manifold. These theorems of Edwards and Quinn classify n-
manifolds among generalized n-manifolds as follows: *A generalized n-manifold* X, n ≥ 5, *is an n-manifold if and only if*
X *satisfies* DDP. On the other hand, the generalized n-manifolds for which DDP fails may deviate from being n-manifolds
in a drastic manner (see [DW] and [S]) and the extent to which
this deviation occurs concerns us. The failure of DDP may
cause enough damage to "the local structure" of a generalized
n-manifold X that X may not contain any proper subset of
dimension >2 which is shape dominated by a polyhedron. More
precisely, we state the following result from [S] as a sample:

THEOREM: For each topological n-manifold M^n without
boundary with n ≥ 5, there exists an uncountable family M^n
of topologically distinct generalized n-manifolds such that
each X in M^n satisfies (a) $X \times E^1$ is homeomorphic to
$M^n \times E^1$, (b) X does not contain any compact proper subset
which is a strongly movable continuum of dimension >2, and
(c) every 1-dimensional compactum embeds in X. This theorem
is an extension of Bing and Borsuk's work [BB]. John Walsh
has informed us that T. Lay (Thesis at U. of Tennessee) has
extended to the Hilbert cube Q a construction of Cantor-
sets which is analogous to Daverman-Edwards' construction for
n-manifolds [DA]. Indeed, by utilizing these Cantor-sets the
content of the theorem, given above, extends immediately by
replacing M^n by Q, i.e., our entire finite-dimensional

program (see [S]) extends to Q. As an additional testi-
monial to the manifestation of the pathology in generalized
n-manifolds one may consult [DW].

The problem at hand is to explore further, i.e., we are
interested in the movable closed subsets of a generalized
n-manifold. We shall show that certain generalized n-manifolds contain many movable proper closed subsets of dimension
≥ 2, see Theorem (5.1.1) for a precise statement.

We have organized this note as follows. Section (2) contains a summary of notation and other conventions. Section
(3) contains a brief study of matters related to the movability or the non-movability of the decomposition space X/A
provided X and A are movable. We are indebted to Ross
Geoghegan for some helpful remarks concerning Section (3).
All in all, Section (3) presents a very brief discussion of
these matters. In Section (4), we observe that generalized
cactoids are movable and we suitably deduce our result, see
Theorem (5.1.1).

2. NOTATION AND TERMINOLOGY

All spaces will be at least separable metric and all maps will
be continuous. By a *decomposition* G of a space X we shall
always mean an upper semicontinuous decomposition of X into
compacta and we let X/G denote the associated decomposition
space. Let N_G and H_G denote the set of all the nondegenerate elements of G and their union, respectively. If A
is a compact subset of a space X, we let X/A denote the
decomposition space of X obtained by identifying A to a
point. For any space X, we let cX and sX denote the
cone over X and the suspension over X, respectively. If
A is a subset of X, we let $X \cup cA$ denote the space obtained by identifying X and cA along A by the identity

map of A. We shall suppress base points whenever this may not
not cause any confusion. We assume familiarity with shape theory
theory [B, DS]. The condition Mittag-Leffler for pro-groups
is abbreviated to ML (cf. [DS]). By Sh(A) and $Sh_o(A)$ we
mean the unpointed and pointed shape of the compactum A. Any
compact, connected, and locally connected metric space is
called a *Peano space*. Given two pointed spaces (X, x_o) and
(Y, y_o), see [B] for a definition of $(X, x_o) \underset{top}{+} (Y, y_o)$:
We suppress base points and denote this space by X + Y.

3. MOVABILITY OF (X,A) AND X/A: RELATED MATTERS

(3.1) Preliminary Remarks. *Throughout this section we consider only pairs of compacta, i.e., a pair* (X,A) *where* X
is a compactum (compact and metric) and A *is a closed subset* X. We use Mardešić-Segal [MS] approach to shape for
pairs of compacta. Our main interest, in this section, is the
movability of X/A where X is a movable compactum and A
is a movable closed subset of X. Since this does not hold in
general (see Theorem (3.1.5)), we seek conditions under which
this is true and explore matters related to this problem. The
details will follow and we begin with the following definition.

(3.1.0) Definition. A pair (X,A) is *Ȟ-admissible* if the
Čech homology sequence of (X,A) is exact with coefficients
in any abelian group.

(3.1.1) A Problem. Give conditions on a pair (X,A)
under which at least one of the following holds:

(a) (X,A) is movable;

(b) X/A is movable; and

(c) (X,A) is Čech Ȟ-admissible.

We shall next discuss some relationship between (a), (b), and (c), as above, and other related matters.

(3.1.2) <u>PROPOSITION</u>: If (X,A) is a movable pair, then X/A is movable.

<u>Proof</u>. Clearly, Sh(X/A) = Sh(X ∪ cA). Since cA is an FAR, our proof is finished by [D].   □

(3.1.3) <u>PROPOSITION</u>: If X/A is movable for a pair (X,A) where X and A are movable, then (X,A) is Ȟ-admissible.

<u>Proof</u>. X/A movable implies that the homology pro-groups of (X,A) satisfy ML. This suffices, since $\underset{\leftarrow}{\text{Lim}}^1$ vanishes in this setting, see [DS,O].

(3.1.4) <u>THEOREM</u>: If (X,A) is not Ȟ-admissible, X movable, and A movable, then X/A is nonmovable.

<u>Proof</u>. Suppose X/A is movable. Choose a nest $(\underline{X},\underline{A})$ $=\{(X_n,A_n)\}_{n=1}^{\infty}$ of polyhedral pairs (A_n is a subpolyhedron of X_n for each n) with intersection (X,A). It is well-known that the singular homology groups $H_*(X_n,A_n;G)$ is naturally isomorphic to $H_*(X_n/A_n;G)$ for any abelian group G and any value of *. This implies that the pro-groups $H_*(\underline{X},\underline{A};G)$ and $H_*(\underline{X/A};G)$ are isomorphic where $\underline{X/A} = \{X_n/A_n\}_n$ is (suitably considered) an ANR-system associated with X/A; and therefore, $H_*(\underline{X},\underline{A};G)$ satisfies ML for every * and any abelian group G. The exact sequence $\cdots \rightarrow H_n(\underline{A};G) \rightarrow H_n(\underline{X};G) \rightarrow H_n(\underline{X},\underline{A};G) \rightarrow H_{n-1}(\underline{A};G) \rightarrow \cdots$ of pro-groups remains exact after passing to the inverse limit since $\underset{\leftarrow}{\text{Lim}}^1$ vanishes in this setting,

see [DS, O] for more details; and hence, (X,A) is Ȟ-admissible which is contrary to our hypothesis.   ☐

(3.1.5) Examples. (a) *There exists a pair* (X,A) *such that* X *is movable,* A *is movable, and* (X,A) *is not* *Ȟ-admissible,* see [O].

(b) We observe that the pair (sX, sA), where (X,A) as above in (a), obtained by suspending (X,A) satisfies: sX *is pointed movable,* sA *is pointed movable, and* (sX,sA) *is not Ȟ-admissible.*

(c) *There exists an Ȟ-admissible pair* (X,A) *such that* X/A *is nonmovable, and hence,* (X,A) *is nonmovable.* Take X to be the continuum of Case and Chamberlain (cf. [B]) and A to be any subcontinuum of trivial shape.

(d) *There exists a nonmovable pair* (X,A) *such that* X/A *is movable, in fact, of trivial shape.* Let X denote the 3-cell and let A denote the Case and Chamberlain continuum inside X. Clearly Sh(X/A) = Sh(sA), and Mardešić (cf. [B]) has shown that sA has trivial shape.

This concludes our brief study of relationships between (a), (b), and (c) of (3.1.1) and we begin the following problem.

(3.2) Problem. Given a pair (X,A) such that X is (pointed) movable and A is (pointed) movable. Find conditions, on the pair (X,A), under which the decomposition space is X/A (pointed) movable.

We emphasize that the movability of X/A depends on the embedding of A in X (this is similar to the movability of the pair (X,A)). The following example illustrates this.

(3.2.0) Example. *There exist two pairs* (X,A) (X,A') *such that: (a)* A *is homeomorphic to* A', *(b)* A *is pointed*

movable,   (c) X *is pointed movable,   (d)* X/A *is movable, and*
(e) X/A' *is nonmovable.* Let   (Y,B)   be any pair satisfying
Y pointed movable, B movable, and (Y,B) is not $\check{H}$-admis-
sible, see Example (3.1.5). Put X = Y + Q and A = B. Let
A' be a subset of Q homeomorphism to A. Now X/A is nonmovable since Sh(X/A) = Sh(Y/A + Q) = Sh(Y/A) and Y/A is
nonmovable by Theorem (3.1.4); and, X/A' is movable since
Sh(X/A') = Sh(X ∪ cA') = Sh(X ∪ cA'/Q) = Sh(X + sA') where X
is pointed movable and sA' is pointed movable, and hence,
X/A' is pointed movable. The pairs (X,A) and (X,A')
satisfy the desired properties.

The following is intended as a brief study of Problem (3.2).

(3.2.1) <u>Definition.</u> A pair of compacta (Y,B) is a *soft-pair* (abbreviation: *s-pair*) if there exists a continuum A
of trivial shape such that B ⊂ A ⊂ Y.

(3.2.2) <u>PROPOSITION:</u> If (X,B) is an s-pair where X is
a pointed movable continuum and B is a compactum whose suspension is pointed movable, then X/B is a pointed movable
continuum (and hence, (X,B) is $\check{H}$-admissible).

<u>Proof.</u> Observe $Sh_o(X/B) = Sh_o(X ∪ cB) = Sh_o(X ∪ cB/A)$,
where A is a continuum of trivial shape with B ⊂ A. Since
$Sh_o(X ∪ cB/A) = Sh_o(X/A + sB)$, our proof is finished (see
[B] for details).   □

(3.2.3) <u>Remark.</u> It is known that the suspension of a compactum B is pointed movable provided: (a) B is movable
(this applies when B is 0-dimensional), or (b) B is a continuum whose Čech cohomology with integral coefficients is
finitely generated [GL].

(3.2.4)   A Theorem of R. H. Bing: Theorem (3.2.5), given
below, is a version of Theorem (19) of Bing [Fund. Math. 36
(1949), 303-318] (see also [WI_2]); however, our proof, which
is included here for completeness, appears to be rather
elementary.

(3.2.5)   THEOREM: If   C   is a 0-dimensional closed subset
of a Peano continuum   P,   then there exists a dendrite (cf.
[WH_1]) D   such that   C ⊂ D ⊂ P.

Proof. Without loss of generality suppose   C   is a Cantor
set.   Let   B   denote the "middle-third" Cantor set in   I.   Let
I_1 denote the closure of the first interval removed, from   I
to construct   B,   let I_2 and I_3 denote the closures of the
two intervals removed from   I   at the second stage, and we
continue in this manner to find I_1, I_2, I_3,⋯ .   Let
$\phi : I \to X$ be a map such that ϕ maps   B   homeomorphically
onto   C,   and ϕ maps each I_n homeomorphically onto a subset A_n of   P.   Put $X = \phi(I)$.   Let   D   be an irreducible
subcontinuum of   X   about   C   (cf. [WI_1], p. 17).   We shall
prove: (a)   D   does not contain any s.c.c. ("simple closed
curve"), and   (b)   D   is l.c. ("locally connected").

By a *segment* in any space   Z   we mean an open subset of   Z
which is homeomorphic to the open interval   (0,1).   Observe
that the sequence of arcs A_1, A_2,⋯   is null.   It follows
that: If there is a sequence $n_1 < n_2 < \cdots$ of indices and
a sequence $\{a_{n_i}\}_{i=1}^{\infty}$ of points, with a_{n_i} in A_{n_i},   converging to a point   x   in   X,   then   x   belongs to   C.   Sup-
pose   D   contains a s.c.c., say,   S.   Let   x   be a point in
(S − C).   It follows that there exists a subset   J   of   (S − C)
containing   x   such that   J   meets finitely many of the arcs
A_1, A_2,⋯ .   We analyze the subsets of   J   where these arcs

meet J and conclude that there is a segment L of D contained in J. This is elementary and we omit details. Observe that (D -L) is arcwise connected since any arc passing through L can be turned around to go through (S -L). This contradicts irreducibility of D and we conclude that D does not contain any s.c.c.

We now prove that D is l.c. Our proof is again by contradiction. Suppose Y is a subcontinuum of D such that D is not l.c. at any point of Y. Choose a point y in (Y -C). Let U be an open neighborhood of x such that U does not meet C. Now there is a continuum of convergence contained in U and hence in the union of arcs A_1, A_2, $\cdots$. This is impossible since these arcs form a null collection. This proves that D is l.c.   □

(3.2.6) <u>COROLLARY</u>: The pair (P,C), and P and C as above in Theorem (3.2.5), is an s-pair.

(3.2.7) <u>COROLLARY</u>: Given P and C as above. Then $Sh_o(P/C) = Sh_o(P/B)$ where B is any closed subset of P having the shape of C.

(3.2.8) <u>COROLLARY</u>: If C is a 0-dimensional subset of a movable Peano continuum P, then P/C is movable.

4. DECOMPOSITIONS OF MANIFOLDS AND MOVABILITY

(4.1) <u>Cactoids</u>. All our 2-manifolds are compact, connected, without boundary, and may not be orientable. A Peano continuum P is defined to be a *cactoid* if each true cyclic element of P is a 2-sphere; see [WH_1]. The following is a classical theorem of R. L. Moore: *A Peano continuum P is a cactoid if and only if there exists a monotone map from the 2-sphere* S^2 *onto* P, *i.e.,* P *is a monotone image of* S^2. We use the terminology of Gmurczyk [G] concerning bouquets. The following proposition, stated here without proof, is straightforward but tedious to prove

(4.1.0) PROPOSITION: Every cactoid has the pointed shape of a bouquet of 2-spheres, i.e., each leaf of the bouquet is a 2-sphere.

(4.1.1) Generalized Cactoids. A Peano continuum P is a *generalized cactoid* if each true cyclic element of P is a 2-manifold and all but finitely many are 2-spheres. Although every cactoid is a generalized cactoid, we have followed the tradition to treat them separately. The following is a theorem of Steenrod and Roberts (cf. [WH$_1$]): *A Peano continuum* P *is a generalized cactoid if and only if* P *is a monotone image of a 2-manifold.*

(4.1.2) PROPOSITION: Every generalized cactoid has the pointed shape of a bouquet of 2-manifolds and all but finitely many are 2-spheres.

Since every bouquet of 2-manifolds is pointed movable [G], the following theorem is immediate.

(4.1.3) THEOREM: All monotone images of 2-manifolds are pointed movable, i.e., every generalized cactoid is pointed movable.

(4.1.4) Remark. Observe that a monotone image of a generalized cactoid is, again, a generalized cactoid and hence pointed movable.

(4.2) **Preliminaries on Decompositions**. Given a decomposition
G of a space X. G will be called *movable, countable, k-dimensional, compact*, or *null* if each element of G is
movable, N_G is countable, the image of H_G is a k-dimensional subset of X/G, H_G is compact, or the diameters of
elements of N_G converge to 0, respectively.

(4.2.0) <u>Decomposition Spaces up to Shape: G Null</u>. Suppose G is a null decomposition into compacta of a metric
space X. Enumerate elements of N_G as $g_1, g_2, \cdots$. Consider the subset X_1 of X × [0,1] obtained as the union of
sets X × {0}, g_1 × [0,1], g_2 × [0,1/2], $\cdots$.
    (A) The collection G_1 of subsets of X_1 whose
$N_{G_1} = \{g_1 \times \{1\}, g_2 \times \{\frac{1}{2}\}, \cdots\}$ is a decomposition of X_1.
(Recall: All our decompositions are upper semicontinuous.)
    (B) The space X_1 is compact provided X is compact.
    (C) The decomposition space $\hat{X} = X_1/G_1$ can be mapped onto
the decomposition space X/G under a map whose point-inverses
are of trivial shape.

The space $\hat{X}$ is obtained by attaching a cone over each
nondegenerate element such that these cones form a null collection; and, one may map each cone over an element g_n onto
its image in X/G. We omit details.

(4.2.1) <u>Decomposition Spaces up to Shape: G compact</u>. In
this case, consider the subset $X_1 = (X \times \{0\}) \cup (H_G \times [0,1])$
of X × [0,1]. The collection G_1 of X_1 with
$N_{G_1} = \{g \times \{1\} : g \in N_G\}$ is clearly a decomposition of X_1
and we form $\hat{X} = X_1/G_1$. Assertions (B) and (C) of (4.2.0)
remain valid without any alteration.

The following is an immediate consequence of a theorem of
Sher (cf. [B]).

(4.2.2) <u>THEOREM</u>: Suppose X is a compact metric space of finite dimension and G is a compact or null decomposition of X such that dim X/G < ∞. Then, in either case, $\hat{X}$ is a compact metric space having the shape of the decomposition space X/G.

(4.3) <u>Some Images Under Light Mappings: A Shape Classification</u>. A decomposition G of a space X is called *light* if each element of G is a 0-dimensional subset of X. (Recall: We only consider decompositions into compacta and all spaces are at least separable metric.) Any space Y homeomorphic to the subspace $\{0, 1, \frac{1}{2}, \frac{1}{3}, \cdots\}$ of [0,1] will be called *simple*. Let A denote the simple subset of $S^2 = \{(x,y,z) : x^2 + y^2 + z^2 = 1\}$ obtained as the closure of the subset $\{(1/2^n, \sqrt{2^n - 1/2^n}, 0) : n = 0, 1, 2, \cdots\}$ of S^2. The decomposition space $M = S^2/A$ will be called *the model space*. The following preliminary proposition justifies this name for M.

(4.3.0) <u>PROPOSITION</u>: If G is a null and light decomposition of S^2 such that H_G is infinite, and H_G is contained in a compact subset of S^2, then $Sh_o(S^2/G) = Sh_o(M)$.

<u>Proof</u>. Choose a disk D such that $H_G \subset D$. Observe $Sh_o(S^2/G) = Sh_o(\hat{S}^2) = Sh_o(\hat{S}^2/D) = Sh_o(S^2 + B)$ where B is a disperse bouquet whose leaves are suspensions over the elements in N_G. Since B is a plane continuum, it follows that $Sh_o(B) = Sh_o(sA)$ (see [B]); and therefore, $Sh_o(S^2 + B) = Sh_o(S^2 + sA) = Sh_o(M)$.   □

(4.3.1) <u>Additional Remarks</u>. Proposition (4.3.0) and related discussions remain valid for arbitrary S^n; but we shall continue, for simplicity, our study of decompositions of s^2 and point out the generalities in the end. The case when H_G is not contained in a compact subset of s^2 requires a separate proof from the one given above. We have preferred to treat these two cases separately (since the proof when H_G is contained in a compact subset of s^2 is simpler and conceptually different).

(4.4)   Attaching Suspensions to a Space.   Suppose sX_1, sX_2, $\cdots$ are disjoint suspensions over compact spaces X_1, X_2, $\cdots$. Suppose X is a space. Let X' denote the disjoint union of X, sX_1, sX_2, $\cdots$. Choose a suspension point from each suspension sX_n and identify it with a point x_n of X and we have sequence x_1, x_2, $\cdots$ of points of X (a point may be repeated in this sequence). We form a set $\hat{\hat{X}}$ as a quotient set of X' under these identifications; and furthermore, we topologize $\hat{\hat{X}}$ so that each of the spaces X, sX_1, sX_2, $\cdots$ is embedded in $\hat{\hat{X}}$ and the sequence of diameters of sX_1, sX_2, $\cdots$ converges to zero. Observe that the space $\hat{\hat{X}}$ depends on, among other things, the choice of the sequence x_1, x_2, $\cdots$.

(4.4.0)   PROPOSITION: Suppose X is a finite dimensional continuum with homotopy type of s^2 and suppose that sX_1, sX_2, $\cdots$ are disjoint suspensions over compact 0-dimensional sets X_1, X_2, $\cdots$. Then, $Sh_o(\hat{\hat{X}}) = Sh_o(M)$, where M is the model space (see (4.3)).

Proof.   We first handle the case when $X = s^2$. Suppose the sequence x_1, x_2, $\cdots$ is dense in X; for otherwise, our proof is the one given for Proposition (4.3.0). Choose a point $x_o \neq x_n$, $n \geq 1$, in X and "blow it up into a disk":

This means that we choose a map $f : X_o \to X$, $X_o = S^2$, such that $f^{-1}(x_o)$ is a closed disk and $f^{-1}(x)$ contains exactly one point for $x \neq x_o$. Consider a sequence of points y_1, $y_2, \cdots$ of X where $\{y_n\} = f^{-1}(x_n)$ for $n = 1,2,\cdots$. Attach a copy sX_n to X_o along y_n, for $n = 1,2,\cdots$, and construct a space $\hat{X}_o$ and a map $\hat{f} : \hat{X}_o \to \hat{X}$ which extends f and $\hat{f}$ maps each suspension homeomorphically to its corresponding copy. Clearly, $\hat{f}$ is cell-like, and therefore, $\hat{\hat{f}}$ is a shape equivalence. Since $Sh_o(\hat{\hat{X}}_o) = Sh_o(M)$ (see proof of Proposition (4.3.0)), our proof is finished when $X = S^2$.

Let $f : X \to S^2$ be a homotopy equivalence with mapping cylinder M_f. Since f is a homotopy equivalence, $Sh_o(M_f) = Sh_o(X)$. We define a space $\hat{M}_f$ by considering that the suspensions are attached along X but disjoint from M_f along the sequence $x_1, x_2, \cdots$. Attach a copy of sX_n to S^2 along $f(x_n)$, $n = 1,2,\cdots$, and construct a space $\hat{S}^2$ such that f extends to map $\hat{f} : \hat{X} \to \hat{S}^2$ which is a homeomorphism when restricted to each suspension. Clearly, $Sh_o(\hat{X}) = Sh_o(\hat{M}_f)$. Now, the cell-like retraction of M_f onto S^2 and the map $\hat{f}$ can be combined to obtain a cell-like mapping $\hat{\hat{M}}_f$ onto $\hat{\hat{S}}^2$. The details are easy and we omit them. This proves that $Sh_o(\hat{X}) = Sh_o(M)$. $\square$

(4.5) <u>The Shape of the Model: Additional Results</u>. The proof of Proposition (4.3.0) depended on the fact that H_G is contained in a compact subset of S^2. We now remove this restriction as follows.

(4.5.0) <u>THEOREM</u>: If G is a null and light decomposition of S^2 such that H_G is infinite, then $Sh_o(S^2/G) = Sh_o(M)$. Nevertheless, if G is any null and light decomposition of S^2, then S^2/G is pointed movable.

Proof. We must handle the case when H_G is dense. Consider $X = S^2 \times B^n$ where B^n is a closed n-ball for n fixed, $n \geq 3$, and identify S^2 with $S^2 \times \{0\}$ (we are not interested in economy concerning the size of n). Let C_1, $C_2, \cdots$ denote an enumeration of elements in N_G. Choose disjoint arcs A_1, $A_2, \cdots$ such that each C_n is contained in A_n and the decomposition G of X into these arcs and singleton sets is shrinkable (a discerning reader may notice our heavy handedness). Consider the space $\hat{X}$ obtained from X by attaching cones to X along C_1, $C_2, \cdots$, see (4.2). Now the decomposition G of X extends to a decomposition $\hat{G}$ of $\hat{X}$ by adding singleton sets. Observe that $\hat{X}/\hat{G}$ is homeomorphic to space $\hat{\hat{X}}$ which is obtained by attaching suspensions, over C_1, $C_2, \cdots$, to X in a suitable manner. Since X has the homotopy type of S^2, we have $Sh_o(\hat{\hat{X}}) = Sh_o(M)$ by Proposition (4.4.0). Observe that if H_G is finite then S^2/G is an ANR. $\square$

(4.6) The Shape of the Model of S^n. Consider the set A described in (4.3) as a subset of S^n, $n \geq 1$, by $A \subset S^1 \subset S^n$. Let $M = M(S^n) = S^n/A$ denote *the model*. Let W denote the Hawaiian ear ring. Our techniques of (4.3) – (4.5) also prove the following.

(4.6.0) THEOREM: If G is a null and light decomposition of S^n with $n \geq 1$ (or a continuum X of trivial shape) such that H_G is infinite, then $Sh_o(S^n/G) = Sh_o(M)$ ($Sh_o(X/G) = Sh_o(W)$). Nevertheless, S^n/G (X/G) is pointed movable for any null and light decomposition of S^n (X).

(4.6.1) <u>Remark</u>. Our techniques can be easily adapted to extend Theorem (4.6.0) much further but we shall pursue this separately because of many technical details involved.

5. MOVABLE COMPACTA IN GENERALIZED n-MANIFOLDS

(5.1) <u>Amalgamations of Some Decompositions</u>. Throughout, let $\pi : M^n \to M^n/G$ denote the projection onto the decomposition space associated with a CE decomposition G of an n-manifold M^n, $n \geq 3$. A CE map $\lambda : M^n/G \to X$ onto some space X will be called an *amalgamation* (Abbreviated: amal.) of π (or G). An amal. $\lambda : M^n/G \to X$ of π is a *null amal.* if the decomposition $\{\xi^{-1}(x)\}_{x \in X}$ of M^n induced by the composite $\xi = \lambda\pi$ is null. The following result (stated here without proof) is communicated to us by R. J. Daverman.

(5.1.0) <u>Existence of Amalgamations</u>. Given $\pi : M^n \to M^n/G$ as above; in addition, suppose G is 0-dimensional. Then there exists a null amal. $\lambda : M^n/G \to X$ of π (such that non-degenerate point-inverses are trees).

(5.1.1) <u>THEOREM</u>: If G is a 0-dimensional CE decomposition of an n-manifold, $n \geq 3$, then the generalized n-manifold M^n/G contains a movable continuum A of dimension ≥ 2 such that either A has the shape of the model M (described in (4.3)) for S^2, or A is an ANR.

<u>Proof</u>. Let $\pi : M^n \to M^n/G$ and $\lambda : M^n/G \to X$ be denoted as in (5.1.0). Put $\xi = \lambda\pi : M^n \to X$. Choose any 2-sphere Σ inside M^n and consider the restriction $\xi| : \Sigma \to \xi(\Sigma)$ of ξ. Factor the map $\xi|$ by the monotone-light factorization to obtain the diagram,

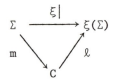

where m is monotone and ℓ is light.   Choose a 2-sphere
Σ_1 in the cactoid and consider $\ell(\Sigma_1)$.   By Theorem (4.5.0),
$\ell(\Sigma_1)$ has the shape of the model M or it is an ANR.   Observe that $A = \lambda^{-1}[\ell(\Sigma_1)]$ and $\ell(\Sigma_1)$ have the same shape
and our proof is finished.   □

The following question is somewhat related to the results
of this note.

(5.1.2) Question.   Does every generalized n-manifold (or
an n-dimensional movable continuum) with $3 \leq n < \infty$ contain
a movable continuum of dimension ≥ 2 other than itself?

(5.1.3) Concluding Remarks.   Question (5.1.2) remains open,
in particular, for a generalized n-manifold constructed by R.
J. Daverman and J. J. Walsh [DW].   Theorem (5.1.1) can be
easily extended to 0-dimensional decompositions which may not
be CE.   We shall not discuss these matters since the details
are lengthy.

REFERENCES

[B]   Borsuk, K., "Theory of Shape." Monografie Matematyczne
        59, Warszawa (1975).

[BB]   Bing, R. H. and K. Borsuk, "A 3-dimensional Absolute
        Retract Which Does Not Contain Any Disk," 54 (1964),
        159-175.

[C_1] Cannon, J. W., "The Recognition Problem: What is a
        Topological Manifold?" Bull. Amer. Math. Soc. 84
        (1978), 832-866.

[C_2] ------, "Taming Codimension-one Generalized Submani-
        folds of S^n." Topology 16 (1977), 323-334.

[D]   Dydak, J., "Movability and the Shape of the Decomposi-
        tion Spaces." Bull. Acad. Polon. Sci. 23 (1975), 57-60.

[DA]   Daverman, R. J., "On the Absence of Tame Disks in Cer-
        tain Wild Cells." Lecture Notes in Math., Springer,
        (1974), 142-145.

[DS]   Dydak, J. and J. Segal, "Shape Theory." Lecture Notes
        in Math. 688, Springer-Verlag, Berlin-Heidelberg,
        New York (1978).

[DW]   Daverman, R. J. and J. J. Walsh, "A Ghastly Generalized
        n-Manifold," preprint.

[E]   Edwards, R. D., "Approximating Certain Cell-like Maps
        by Homeomorphisms." Notices Amer. Math. Soc. 24 (1977),
        A649. Abstract #751-G5.

[G]   Gmurczyk, "On Bouquets." Fund. Math. 93 (1976), 161-
        179.

[GL]   Geoghegan, R. and R. C. Lacher, "Compacta with the
        Shape of Finite Complexes." Fund. Math. 92 (1976),
        25-28.

[L]   Lacher, R. C., "Cell-like Mappings and Their General-
        izations." Bull. Amer. Math. Soc. 83 (1977), 495-552.

[M]   Mardešić, S., "On the Shape of the Quotient Space
         S^n/A." <u>Bull. Acad. Polon. Sci.</u> 19 (1971), 623–629.

[MS]   Mardešić, S. and J. Segal, "Shapes of Compacta and
         ANR-Systems." <u>Fund. Math.</u> 72 (1971), 41–59.

[O]   Overton, R., "Čech Homology for Movable Compacta."
         <u>Fund. Math.</u> 77 (1973), 241–251.

[Q]   Quinn, F., "Ends of Maps, I." <u>Ann. of Math.</u> 110 (1979),
         275–331.

[S]   Singh, S., "Generalized Manifolds (ANR's and AR's) and
         Null Decompositions of Manifolds." <u>Fund. Math.</u> (to
         appear).

[WH$_1$] Whyburn, G. T., "Analytic Topology." <u>Amer. Math. Soc.</u>
         <u>Colloq. Publ.</u>, Vol. 28, Providence, R.I. (1942).

[WH$_2$] ------, "Concerning the Proposition that every Closed,
         Compact, and Totally Disconnected Set of Points is a
         Subset of an Arc." <u>Fund. Math.</u> 18 (1932), 47–60.

[WI$_1$] Wilder, R. L., "Topology of Manifolds." <u>Amer. Math.</u>
         <u>Soc. Colloq. Publ.</u>, Vol. 32, Providence, R.I. (1949).

[WI$_2$] ------, "Monotone Mappings of Manifolds, II." <u>Michigan</u>
         <u>Math. J.</u> 5 (1958), 19–23.

On Almost Continuous Functions

Marwan M. Awartani and Samir A. Khabbaz

In this paper we classify functions in the class D of almost
continuous functions, cf. Definition 1, by obtaining homotopy
invariants of their graphs. We also construct uncountably
many homotopy classes of graphs of functions in D.

Definition 1. Let D denote the class of all functions
$f : I \to I$, I being the closed unit interval [0,1] having the
following three properties: (i) f is discontinuous only at
0, (ii) f(0) = 0, (iii) the graph of f is a connected
subspace of the Euclidean plane R^2.

Notation. Throughout this paper the following notations
will be used: For $f \in D$, let G_f denote the graph of f;
and let o_f denote the point (0,0) in G_f. If $Y \subseteq X \subseteq G_f$,
then d*(X) denotes the diameter of X and C(X) denotes
the set of connected components of X. If $L \in C(Y)$, then
$car_X(L)$ is the element of C(X) that contains L. If p
and q are two points in G_f distinct from o_f, then
$[p,q]_f$ denotes the compact segment of G_f determined by p
and q. Finally let ω denote the set of positive integers.
The word map denotes a continuous function.

Definition 2. Let f and g belong to D. A map
$\alpha : G_f \to G_g$ is called *nontrivial* if $\alpha(o_f) = o_g$.

Remark. Let f and g be in D. It is easy to verify
that if $\alpha : G_f \to G_g$, $\beta : G_g \to G_f$ are two maps with
$(\beta o \alpha) \simeq 1_{G_f}$, where 1_{G_f} is the identity map on G_f and

where $\simeq$ indicates homotopy, then both α and β must be nontrivial.

Definition 3. Let $f \in D$ and let $\{K_i\}$ be a sequence of subsets of G_f. Then $\{K_i\}$ is called *essential* if the sequence $\{d*(K_i)\}$ does not cluster at 0. If $\{d*(K_i)\} \to 0$, then $\{K_i\}$ is called *inessential*. Moreover, if $\{P_i\}$ is a sequence of points in $G_f \backslash o_f$ converging to o_f, then $\{P_i\}$ is called *essential* if for each subsequence $\{P_{i_j}\}$ of $\{P_i\}$, the sequence $\{[P_{i_j}, P_{i_{j+1}}]_f\}$ is essential. Here $G_f \backslash o_f = \{x \in G_f | x \neq o_f\}$.

Definition 4. Let f and g be functions in class D and let $\alpha, \beta : G_f \to G_g$ be two nontrivial maps. Then α and β are called *nonseparated* if for each sequence $\{P_i\}$ of points in G_f which converges to o_f, the sequence $\{[\alpha(P_i), \beta(P_i)]_g\}$ is inessential. If α and β are not nonseparated we call them *separated*.

It follows from the above definition that if α and β are separated maps from G_f to G_g, then there exists a sequence $\{P_i\}$ of points in G_f which converges to o_f, and a sequence $\{q_i\}$ of points in G_g which is bounded away from o_g such that for each $i \in \omega$, $q_i \in [\alpha(P_i), \beta(P_i)]_g$.

Notation 5. Let f and g be functions in class D and let $\alpha : G_f \to G_g$ be a nontrivial map. Then let η_α denote the function $\Pi_1 \circ \alpha \circ i : I \to I$, where $i : I \to G_f$ is the function given by $i(t) = (t, f(t))$, and $\Pi_1(x,y) = x$.

THEOREM 6: Let f and g be functions in class D and suppose that $\alpha, \beta : G_f \to G_g$ are two nontrivial maps. Then α and β are homotopic if and only if they are nonseparated.

Proof. We first prove that if α and β are homotopic, then they are nonseparated. Suppose that α and β are separated and that $H : G_f \times I \to G_g$ is a homotopy between them. Since α and β are separated, there exists a sequence $\{P_i\}$ of points in G_f which converges to o_f, and a sequence $\{q_i\}$ of points in G_g which is bounded away from o_g, such that for each $i \in \omega$, $q_i \in [\alpha(P_i), \beta(P_i)]_g$. Now for each $i \in \omega$, let t_i be a point in I such that $H(P_i, t_i) = q_i$. Since I is compact, $\{t_i\}$ contains a subsequence $\{t_{i_j}\}$ which converges to some point $t^* \in I$. Since $\alpha(o_f) = \beta(o_f) = o_g$, it follows that $H(o_f, t) = o_g$ for all $t \in I$. By continuity of H, we have $o_g = H(o_f, t^*) =$
$$= H(\lim_{j \to \infty} P_{i_j}, \lim_{j \to \infty} t_{i_j}) = \lim_{j \to \infty} (H(P_{i_j}, t_{i_j})) = \lim_{i \to \infty} q_{i_j}.$$ This is a contradiction since $\{q_i\}$ is bounded away from o_g.

Conversely: let $\alpha, \beta : G_f \to G_g$ be two nonseparated maps and let $\eta_\alpha, \eta_\beta : I \to I$ be as in Notation 5. Define the function $H : G_f \times I \to G_g$ by $H((x, f(x)), t) = [(1-t)\eta_\alpha(x) + t\eta_\beta(x), g((1-t)\eta_\alpha(x) + t\eta_\beta(x))]$. It can be easily shown that H is a homotopy map between α and β.

Definition 7. Let $f \in D$ and let G be a neighborhood of o_f. Let $\{K_i\}$ be a sequence of distinct elements of $C(G)$. We say that $\{K_i\}$ *admits a positive integer* n, if there exists an essential sequence $\{P_i\}$ of points in G_f such that $|\{P_i\} \cap K_j|$, the number of elements P_i which are in K_j, is equal to n for all but finitely many $j \in \omega$. Moreover, if $\{K_i\}$ admits n and no subsequence of $\{K_i\}$ admits $n+1$, then $\{K_i\}$ is called *n-fold*.

Definition 8. Let $f \in D$ and let G be a neighborhood of o_f. We say that $C(G)$ is *convergence-nontrivial on some integer* n, in symbols $C(G)$ is *c-nontrivial on* n, if for each neighborhood N of o_f with $N \subseteq G$, $C(N)$ contains an n-fold sequence $\{K_i\}$ such that $\{car_G(K_i)\}$ is also n-fold. The *c-frequency of* G_f is defined to be the set of positive integers n such that there exists a neighborhood N of o_f with $C(N)$ c-nontrivial on n.

Example: the following graph is of c-frequency $\{1\}$.

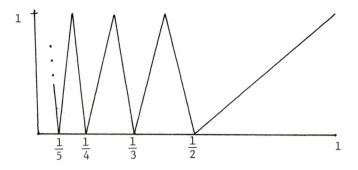

THEOREM 10:   The c-frequency of a graph is a homotopy invariant.

The proof of Theorem 10 follows from Lemma 14. First we need the following:

LEMMA 11:   Let $f \in D$ and let $h : G_f \to G_f$ be a map homotopic to 1_{G_f}. If H is any neighborhood of o_f, then $G = \bigcup_{K \in C(H)} (h^{-1}(K) \cap K)$ is also a neighborhood of o_f, having the property that $h(U) \subseteq car_H h(U)$ for each $U \in C(G)$.

Proof. It clearly suffices to show that G is a neighborhood of o_f. Suppose that $\{P_i\}$ is a sequence of points in G_f which converges to o_f such that $\{P_i\} \cap G = \phi$. We may assume that $\{P_i\} \subseteq H$. For each $i \in \omega$, let $K_i = car_H (P_i)$.

Since $\{P_i\} \cap G = \phi$, it follows that $h(P_i) \notin K_i$ for all $i \in \omega$. Thus the sequence $\{[P_i, h(P_i)]_f\}$ is essential, and h and 1_{G_f} are separated. This contradicts Theorem 6.

LEMMA 12: Let f and g be elements of D, and let $\alpha : G_f \to G_g$, $\beta : G_g \to G_f$ be two maps such that $(\beta \circ \alpha) \simeq 1_{G_f}$. If $\{P_i\}$ is an essential sequence of points in G_f, then $\{\alpha(P_i)\}$ is an essential sequence of points in G_g.

Proof. Set $h = (\beta \circ \alpha)$. Suppose that $\{s_i\}$ and $\{t_i\}$ are two subsequences of $\{P_i\}$ such that $\{[\alpha(s_i), \alpha(t_i)]_f\}$ is inessential. Since both $\{s_i\}$ and $\{t_i\}$ converge to o_f, the continuity of β implies that $\{[h(s_i), h(t_i)]_f\}$ is inessential.

For each $i \in \omega$, we have $d*[s_i, t_i]_f \leq d*[s_i, h(s_i)]_f +$ $+ d*[h(s_i), h(t_i)]_f + d*[h(t_i), t_i]_f$. Since $h \simeq 1_{G_f}$, Theorem 6 implies that $\lim\limits_{i \to \infty} d*[h(s_i), s_i]_f = \lim\limits_{i \to \infty} d*[h(t_i), t_i]_f = 0$. This and the fact that $\{[h(s_i), h(t_i)]_f\}$ is inessential implies that $\{[s_i, t_i]_f\}$ is inessential. This is a contradiction.

COROLLARY 13: Let f and g be elements of D and let $\alpha : G_f \to G_g$, $\beta : G_g \to G_f$ be two maps such that $(\beta \circ \alpha) \simeq 1_{G_f}$. Suppose that G and H are neighborhoods of o_f and o_g respectively with $\alpha(G) \subseteq H$. Let $\{K_i\}$ be a sequence in $C(G)$ such that $\{car_H(\alpha(K_i))\}$ consists of distinct elements. If $\{K_i\}$ admits n, then $\{car_H(\alpha(K_i))\}$ also admits n.

Proof. Let $\{P_i\}$ be an essential sequence of points in G_f with $|\{P_i\} \cap K_j| = n$ for all but finitely many $j \in \omega$. Lemma 12, implies that $\{\alpha(P_i)\}$ is essential. Moreover,

$|\{\alpha(P_i)\} \cap car_H(\alpha(K_j))| = n$ for all but finitely many $j \in \omega$.
Suppose this is not true. Then $\alpha | (K_j \cap \{\alpha(P_i)\})$ is not
$1-1$ for infinitely many $j \in \omega$. Hence, there exists subsequences $\{s_{i_j}\}$ and $\{t_{i_j}\}$ of $\{P_i\}$ with $s_{i_j} \in K_{i_j}$ and
$t_{i_j} \in K_{i_j}$ for all $j \in \omega$, such that $\alpha(s_{i_j}) = \alpha(t_{i_j})$. The
sequence $\{[\alpha(s_{i_j}), \alpha(t_{i_j})]_g\}$ is obviously inessential, contradicting Lemma 12 and the essentiality of the sequence
$\{P_i\}$.

LEMMA 14: Let f and g be elements of D and let
$\alpha : G_f \to G_g$, $\beta : G_g \to G_f$ be homotopy inverses. Suppose that
H is a neighborhood of o_f such that $C(H)$ is c-nontrivial
on $n \in \omega$. Then $C(\beta^{-1}(H))$ is also c-nontrivial on n.

Proof. Set $(\beta \circ \alpha) = h$; $G = \bigcup_{K \in C(H)} (K \cap h^{-1}(K))$,
$T = \beta^{-1}(H)$, and let T' be any neighborhood of o_g contained in T. Set $G' = \alpha^{-1}(T') \cap G$. By Lemma 11, G and
G' are both neighborhoods of o_f.

Since $C(H)$ is c-nontrivial on n, $C(G')$ contains an
n-fold sequence $\{K_i\}$ such that $\{car_H(K_i)\}$ is also n-fold.
It suffices to prove that both $\{car_{T'}(\alpha(K_i))\}$ and
$\{car_T(car_{T'}(\alpha(K_i)))\}$ are n-fold. Since $h(K_i) \subseteq car_H(K_i)$,
it follows that the sequence $\{car_{T'}(\alpha(K_i))\}$ consists of distinct elements. Since $(\beta \circ \alpha) \simeq 1_{G_f}$, it follows from Corollary 13 that $\{car_{T'}(\alpha(K_i))\}$ admits n. Suppose that
$\{car_{T'}(\alpha(K_i))\}$ admits $n+1$. Since $(\alpha \circ \beta) \simeq 1_{G_g}$, it follows
from Corollary 13 again that $\{car_H(\beta(car_{T'}(\alpha(K_i))))\}$ admits
$n+1$. However, for each $i \in \omega$, $car_H(\beta(car_{T'}(\alpha(K_i)))) =$
$= car_H(h(K_i)) = car_H(K_i)$. This implies that $\{car_H(K_i)\}$
admits $n+1$, contradicting the fact that $\{car_H(K_i)\}$ is
n-fold. It follows that $\{car_{T'}(\alpha(K_i))\}$ is n-fold. Since
T' was an arbitrary neighborhood contained in T, it follows

that $\{car_T(\alpha(K_i))\}$ is also n-fold. The result now follows
since $car_T(\alpha(K_i)) = car_T(car_{T'}(\alpha(K_i)))$ for all $i \in \omega$.
This concludes the proof of the lemma and therefore also of
Theorem 10.

The next theorem shows the existence of uncountably many
homotopy classes of graphs of function in the class D.

THEOREM 15: For any set A of positive integers that contains 1, there exists a function $f \in D$ such that G_f is
of c-frequency A.

The condition that A contains 1 is necessary due to
the following lemma whose proof is left to the reader.

LEMMA 16: If the c-frequency A of a graph of a function
in the class D is nonempty, then $1 \in A$.

To prove Theorem 15, we need the following:

Definition 17. Let n and m be two positive integers
with $m > 1$. An $(n, \frac{1}{m})$ block is a function
$g : [a,b] \rightarrow [0,1]$, $[a,b] \subseteq (0,1]$, with the property that
there exists a sequence of $2n + 1$ points $a_i \in [a,b]$ with
$a_0 = a < a_1 < a_2 < \cdots < a_{2n-1} < a_{2n} = b$ such that the following two conditions hold: (i) $g(a_0) = g(a_{2n}) = 1$;
$g(a_i) = 0$ for i odd; $g(a_i) = \frac{1}{m}$ for i even, $i \neq 0$.
(ii) The graph G_g is obtained by joining $(a_i, g(a_i))$ to
$(a_{i+1}, g(a_{i+1}))$ by a straight line segment, $i = 0, \cdots, 2n - 1$.

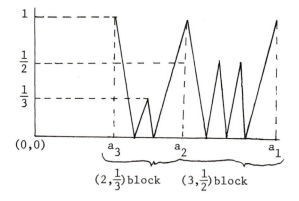

Proof of Theorem 15. We now describe how to construct a function $f \in D$ whose graph is of c-frequency A.

Let $a_1, a_2, a_3, \cdots$ be a decreasing sequence of points in $(0,1]$ such that $a_1 = 1$ and $\{a_i\} \to 0$. Let $f : [0,1] \to [0,1]$ be any function having the following three properties:
(i) $f(0) = 0$; (ii) $f|[a_{i+1}, a_i]$ is an $(n, \frac{1}{m})$ block for some $n \in A$; (iii) for each $n \in A$ and each $m \in \omega$, $m > 1$, there exists infinitely many $i \in \omega$, for which $f|[a_{i+1}, a_i]$ is an $(n, \frac{1}{m})$ block.

The verification that G_f is of c-frequency A is left to the reader.

On the Problems Related to Linear Homeomorphisms, Embeddings, and Isotopies

Robert Connelly, David W. Henderson,
Chung-wu Ho, and Michael Starbird

I. INTRODUCTION

In this paper, we shall make a list of unsolved problems related to the linear embeddings and linear isotopies of a finite
complex into a euclidean space. Let C be a finite complex
with a fixed triangulation T. By a *linear embedding* of (C,T)
into E^n, we mean an embedding of C into E^n which is
simplexwise linear with respect to the triangulation T. Two
linear embeddings f and g of (C,T) into E^n are said to
be *linearly isotopic* if there is an isotopy h_t between f
and g such that for each t, h_t is a linear embedding of
(C,T) into E^n. The questions that we ask are basically of
the following types: When can two linear embeddings of (C,T)
into E^n be linearly isotopic? What are the homotopy groups
of certain spaces of linear embeddings of a triangulated complex into E^n?

There are two main reasons for studying these questions.
First, these are natural questions to ask if one wishes to
study the linear structure of a space or the complexity of a
triangulation. Note that by means of the "Alexander trick,"
many isotopy problems become trivial in the PL-category. For
instance, one can easily show that if D is a polyhedral
n-cell in E^n, then the space of all PL-embeddings of D into
E^n, which are pointwise fixed on $Bd(D)$, is contractible.
However, in carrying out the contraction, what one does is,
instead of untangle the image of each embedding f back to
the identity map, simply letting the image of f be shrunk to
a point to achieve the identity map. Thus, the difficulty on

the triangulation caused by a linear embedding is not resolved but merely made small and ignored. It is a bit like doing knot theory by pulling a knot to a straight line. But in the linear category, we insist on doing each deformation with respect to a *fixed* triangulation in the entire process. Thus, in deforming a linear embedding of a triangulated n-cell in E^n back to the identity, we have to move the images of the vertices of the triangulation back to their original positions.

The second reason for studying these spaces is their close relationship to the *smoothing problem* in differential topology, the problem of determining the existence and the uniqueness of differentiable structures on a triangulated manifold. In fact, many problems on the space of linear embeddings were first studied in the context of smoothing problems (see [3], [14], [15], [21] and [22]). The homotopy groups of some of these spaces will give immediate information to the smoothing problem (see [5], [14], [21] and [22]).

In the following, we shall divide the problems into a few different groups. In each group, we first give the basic definitions, then try to make a few explanatory comments, and finally delineate some of the past results.

II. THE SPACES $L(D^n, K)$

Let D^n be a polyhedral n-dimensional cell in E^n and K be a rectilinear triangulation of D^n. By a linear homeomorphism of (D^n, K), we mean a homeomorphism $f : D^n \to D^n$ such that f is pointwise fixed on $Bd(D^n)$ and the restriction of f to each simplex σ of K is linear on σ. Note that each such f is completely determined by its image of the vertices of K. We shall let $L(D^n, K)$ be the space of all the linear homeomorphisms of (D^n, K) under the sup-metric. It can be shown that each $L(D^n, K)$ is homeomorphic to an open subset of the euclidean space E^{nk}, where k is the number of the vertices of K that lie in the interior of D^n. In this section, an isotopy between two elements of $L(D^n,K)$ will always be assumed to be pointwise fixed on $Bd(D^n)$. Thus, a linear isotopy between two elements f and g in $L(D^n,K)$ corresponds to a path in the space $L(D^n,K)$ between f and g.

The spaces $L(D^n,K)$ were first studied by S. S. Cairns in his work on the smoothing problems. He showed in 1944 that if D is a geometric 2-simplex in E^2 and K is a triangulation which has no other vertices on $Bd(D)$ except the three vertices of the 2-simplex, then the space $L(D,K)$ is pathwise connected [3], [4]. In 1973, C. W. Ho showed that the space $L(D,K)$ is simply connected [9]. He also studied the case when D^2 is a convex polyhedral disk in E^2 and showed that given two elements f and g in $L(D^2,K)$, there exists a subdivision K' of K such that f and g can be connected by a path in the space $L(D^2,K')$ [11]. This result was improved by R. H. Bing and M. Starbird. They showed, among other things, that if D^2 is a convex polyhedral disk in E^2, or if D^2 is star-shaped and the triangulation K contains no spanning edge on D^2, then the space $L(D^2,K)$ is pathwise connected. No subdivision of K is necessary [1]. On the other hand, if D^2 is not convex, the space $L(D^2,K)$ is in general not pathwise connected [1, Example 4.1]. However, two elements of $L(D^2,K)$ can still be deformed into each other if a subdivision of K is allowed [2]. Recently, R. Connelly and D. Henderson have announced that if D^2 is a convex polyhedral disk in E^2, then the space $L(D^2,K)$ is contractible [7]. This result clears up all the homotopy questions concerning the space $L(D^2,K)$ for a convex disk D^2.

For the 3-dimensional polyhedral cells D^3, M. Starbird showed that the space $L(D^3,K)$ in general is not pathwise

connected even when D^3 is a 3-simplex [17]. However, for a
general triangulated polyhedral 3-cell (D^3, K), and for any
two elements f and g in $L(D^3, K)$, there always exists a
subdivision K' of K such that f and g can be connected
by a path in $L(D^3, K')$ [14,20]. Very little is known in higher
dimensional cases except for a result of Ho that for any n,
and any rectilinear triangulation K of a polyhedral n-cell
D^n such that K has only 2 interior vertices, the space
$L(D^n, K)$ is always contractible [12].

Question 1. Let D^2 be a convex polyhedral disk in E^2.
Let $f \in L(D^2, K)$ be given such that $\sup\{\| f(x) - x\| : x \in D^2\}$
$= \varepsilon_0$. For each $\varepsilon > \varepsilon_0$, is it always possible to find a
linear isotopy H_t^ε between f and the identity map in
$L(D^2, K)$ such that for each t and each x, $\| H_t^\varepsilon(x) - x\| < \varepsilon$?
What if a subdivision of K is allowed for the isotopy?

Question 2. Let D^2 be a not necessarily convex polyhedral
disk in E^2. Is there an $\varepsilon > 0$ such that for any triangulation K of D^2 of mesh less than ε, and any $f \in L(D^2, K)$,
there is a linear isotopy in $L(D^2, K)$ carrying f into the
identity map?

Question 3. Let D^2 be a not necessarily convex polyhedral
disk in E^2, and K be a rectilinear triangulation of D^2.
For each loop Γ in $L(D^2, K)$, does there exist a subdivision
K' of K such that Γ can be shrunk to a point in the space
$L(D^2, K')$?

Question 4. Let D^2 and K be as in Question 3, does
there always exist a subdivision K' of K such that the
space $L(D^2, K')$ is contractible?

Question 5. Given a triangulated convex polyhedral 3-cell
(D^3,K) in E^3, is it always possible to find a subdivision
K' of K such that the space $L(D^3,K')$ is simply connected?
Contractible? What if D^3 is a geometric 3-simplex in E^3?

Question 6. Does Starbird's example [17] exist in E^n for
each $n \geq 4$?

Question 7. For each $n \geq 2$, is each component of the
space $L(D^n,K)$ simply connected? Contractible? What if D^n
is a geometric n-simplex in E^n?

III. LINEAR ISOTOPY AND LINEAR MOVES

Let C be a finite complex with a triangulation K. As defined in §I, a linear isotopy between two linear embeddings f
and g of (C,K) into E^n is an isotopy h_t between f and
g such that for each t, h_t is a linear embedding of (C,K)
into E^n. Note that each linear isotopy between f and g is
obtained by moving the images of the vertices of K under f
continuously to the corresponding images under g (with the
motion being extended linearly onto each simplex of K) such
that no simplex of K is allowed to collapse in the process of
moving the vertices. In constructing the linear isotopies some-
times it is more convenient to consider a special kind of
isotopy, called *linear moves*. These are the linear isotopies
that move the image of only one vertex of K. It is easy to
prove that each linear isotopy can always be achieved by a
finite sequence of linear moves (see for instance [11, §4]).

In 1944, S. S. Cairns showed that if (C,T) is a finite
triangulated complex and if f, g are two linear embeddings
of (C,T) into E^2 such that there is an orientation-preserving homeomorphism h of E^2 such that $h \circ f = g$, then

there is a linear isotopy between f and g [3], [4]. Recently, M. Starbird showed that the corresponding theorem fails
for linear embeddings in E^3 [17], however, such a linear
isotopy exists if we allow a subdivision of the triangulation
T.

Question 8.   What about the higher dimensional cases for
Cairns-Starbird result?

Question 9.   Let K be an unknotted polygonal knot in E^3.
Is it always possible to deform K into a triangle by a sequence of linear moves on the vertices of K?

To answer this question, one might want to consider first
the following two questions.

Question 10.   Given an unknotted polygonal knot K in E^3,
is it always possible to find a 2-dimensional plane P in E^3
and a linear move m on a vertex of K such that the projection of K onto P after the move has a fewer number of
crossings than the projection before the move?

Question 11.   Let f be a linear embedding of a triangulated 2-sphere into E^3.   Can the image be pushed onto the
boundary of a tetrahedron by a sequence of linear moves?

In the following, we shall consider instead of individual
linear embeddings, the space of all the linear embeddings of a
triangulated complex.

Question 12.   Given a polygonal unknotted spanning arc A
of a cube I^3, is the space of all the unknotted linear embeddings of A into I^3, fixed on the ends of A, contractible?

Question 13. Given a triangulated spanning polyhedral
2-disk K in the cube I^3, is the space of all the linear
embeddings of K into I^3, fixed on the boundary of K,
contractible?

Note: An affirmative answer to either question 13 or 14
will apparently lead to a proof of the Smale Conjecture that
the space of all diffeomorphisms of a 3-dimensional ball B^3
onto itself, which are fixed on the boundary of B^3, is contractible. For a brief description of the Smale Conjecture
see [8].

IV. SOME OTHER SPACES OF LINEAR HOMEOMORPHISMS

Question 14. Let M be a 2-dimensional manifold with a
Riemannian metric of constant curvature. By a *geodesic tri-*
angulation of M, we mean a triangulation K of M, whose
vertices are so distributed on M that each 1-simplex $\langle v_i, v_j \rangle$
of K is a geodesic arc of the manifold between v_i and v_j.
Now, let L(M,K) be the space under the sup-metric of all the
homeomorphisms of M onto M which take geodesic triangles
of K onto geodesic triangles in M. Is the space L(M,K)
retractible to the subspace consisting of the isometries of M?

Note: Connelly and Henderson's techniques [7] seem to show
that $L(T^2,K)$ retracts to the isometries of T^2, where K
is a geodesic triangulation of the flat torus T^2.

Question 15. One may also consider the space L(M,K,σ) of
all the homeomorphisms in L(M,K) which fix a given 2-simplex
σ of K, and ask what are its homotopy properties. Of particular interest is the case when M is the 2-sphere S^2.

Note: S. S. Cairns proved that the space $L(S^2,K,\sigma)$ is pathwise connected [3]. As a consequence of this result, he showed that every 4-dimensional triangulated manifold has a differentiable structure [5]. However, nothing is known about the homotopy groups of $L(S^2,K,\sigma)$.

V.   THE SPACE OF ALL THE LINEAR HOMEOMORPHISMS OF AN n-CELL

Let D^n be a polyhedral n-cell in E^n, and $\mathcal{L}(D^n)$ be the union of the spaces $L(D^n,K)$'s for all the possible rectilinear triangulations K of D^n. Note that $\mathcal{L}(D^n)$ is a directed family of topological spaces over the directed system of all the rectilinear triangulations of D^n. Similarly, for each $n \geq 0$, the homotopy groups $\pi_n(L(D^n,K),\mathrm{id})$ for different triangulations K of D^n also form a directed family of groups. We shall define the nth homotopy group of the space $\mathcal{L}(D^n)$ as the directed limit of the groups $\pi_n(L(D^n,K),\mathrm{id})$ with respect to the directed system of all the triangulations K of D^n.

The homotopy groups of the spaces $\mathcal{L}(D^n)$ turned out to be of interest in the smoothing theory (see [21], [14] and [16]). By the work of Cairns [3], [4], the space $\mathcal{L}(D^2)$ is pathwise connected if D^2 is a 2-simplex in E^2. C. W. Ho showed in 1973 that the same space is simply connected [10]. However, by the recent work of Connelly and Henderson mentioned above, all the homotopy groups of $\mathcal{L}(D^2)$ are trivial if D^2 is a convex polyhedral 2-disk in E^2 [7].

For the 3-dimensional case, by a recent result of Starbird [20], we can conclude that $\pi_0(\mathcal{L}(D^3)) = 0$ if D^3 is an arbitrary polyhedral 3-cell in E^3 (the same result can be deduced from the work of Kuiper [14] and Cerf [6] if D^3 is a 3-simplex in E^3). On the other hand, using Milnor's 7-dimensional sphere, Kuiper showed that at least one of the groups

$\pi_3(\mathcal{L}(D^3))$, $\pi_2(\mathcal{L}(D^4))$, $\pi_1(\mathcal{L}(D^5))$ and $\pi_0(\mathcal{L}(D^6))$ is non-trivial, where for each n, D^n is a geometric n-simplex in E^n [14].

Question 16. Determine which of the homotopy groups in Kuiper's result are non-trivial and what these groups are.

Question 17. Let D^n be a polyhedral n-cell in E^n. Then the set $\mathcal{L}(D^n)$ has a group structure under the composition of maps. Is this group a simple group?

Question 18. Let D^n be a polyhedral n-cell in E^n. The set $\mathcal{L}(D^n) = \underset{k}{\cup} L(D^n,K)$ can be considered as the direct limit of the topological spaces $L(D^n,K)$'s. Are the homotopy groups of $\mathcal{L}(D^n)$ under this direct limit topology the same as the groups defined above?

More generally, let $\{(X_\alpha, x_\alpha)\}_{\alpha \in A}$ be a directed family of topological spaces with base point such that the index set is uncountable. Under what condition is it true that

$$\pi_k\left(\underset{\alpha \in A}{\underrightarrow{\lim}}\ (X_\alpha, x_\alpha)\right) \cong \underset{\alpha \in A}{\underrightarrow{\lim}}\ \pi_k(X_\alpha, x_\alpha)\quad ?$$

Note: It is easy to prove that if each compact subset of $\underset{\alpha \in A}{\underrightarrow{\lim}}\ (X_\alpha, x_\alpha)$ is contained in some (X_α, x_α) then the above isomorphism exists. Hence, direct limit functor and the homotopy functor commute if the index set is countable and the spaces (X_α, x_α) are reasonably nice. However, it is not clear if the index set A is uncountable.

Question 19. Is $\mathcal{L}(D^n)$ a topological group if it is equipped with the direct limit topology?

REFERENCES

[1] Bing, R. H. and Starbird, M., "Linear isotopies in E^2,"
     Trans. Amer. Math. Soc. 237 (1978), 205-222.

[2] ------, "Super triangulations," Pacific J. Math. 74 (1978),
     307-325.

[3] Cairns, S. S., "Isotopic deformations of geodesic com-
     plexes on the 2-sphere and plane," Ann. of Math. (2) 45
     (1944), 207-217.

[4] ------, "Deformations of plane rectilinear complexes,"
     Amer. Math. Monthly 51 (1944), 247-252.

[5] ------, "Homeomorphisms between topological manifolds and
     analytic manifolds," Ann. of Math. (2) 41 (1940), 796-808.

[6] Cerf, J., "Sur les difféomorphismes de la sphère de
     dimension trois ($\Gamma_4 = 0$)," Lecture Notes in Math. 53,
     Springer Verlag, 1968.

[7] Connelly, R. and Henderson, D. W., "Contracting simplex-
     wise linear homeomorphisms with springs and bumpers,"
     to appear.

[8] Hatcher, A. E., "Linearization in 3-dimensional topology,"
     preprint.

[9] Ho, C. W., "On certain homotopy properties of some spaces
     of linear and piecewise linear homeomorphisms, I." Trans.
     Amer. Math. Soc. 181 (1973), 213-233.

[10] ------, "On certain homotopy properties of some spaces
     of linear and piecewise linear homeomorphisms, II."
     Trans. Amer. Math. Soc. 181 (1973), 235-243.

[11] ------, "Deforming P.L. homeomorphisms on a convex
     polygonal 2-disk," Pacific J. Math. 55 (1974), 427-439.

[12] ------, "On the space of the linear homeomorphisms of a
     polyhedral n-cell with two interior vertices," Math. Ann.
     243 (1979), 227-236.

[13]   ------, "On the extendability of a linear embedding of
       the boundary of a triangulated n-cell to an embedding of
       the n-cell," to appear.

[14]   Kuiper, N. H., "On the smoothings of triangulated and
       combinatorial manifolds," Differential and Combinatorial
       Topology, Princeton University Press, Princeton, N.J.,
       1965, pp. 3-22.

[15]   Munkies, J., "Differentiable isotopies on the 2-sphere,"
       Mich. Math. J. 7 (1960), 193-197.

[16]   ------, "Concordance of differentiable structures --two
       approaches," Mich. Math. J. 14 (1967), 183-191.

[17]   Starbird, M., "A complex which cannot be pushed around
       in E^3," Proc. Amer. Math. Soc. 63 (1977), 363-367.

[18]   ------, "Linear isotopies in E^3," Proc. Amer. Math. Soc.
       65 (1977), 342-346.

[19]   ------, "Flexible regular neighborhoods for complexes in
       E^3," Top. Proceedings 2 (1977), 593-619.

[20]   ------, "The Alexander linear isotopy theorem in E^3,"
       these proceedings.

[21]   Thom, R., "Des variétés triangulées aux variétés dif-
       férentiables," Proc. Int. Congr. Math., Edinburgh, 1958,
       pp. 248-255.

[22]   Whitehead, J. H. C., "Manifolds with transverse fields in
       euclidean space," Ann. of Math. 73 (1961), 154-212.

Simplicial Complexes Homeomorphic to Proper Self-Subsets Have Free Faces

David W. Henderson

If K is a simplicial complex, then an n-simplex $\sigma^n \subset K$ is a *free face* of K if σ^n is the face of exactly one $(n+1)$-simplex, τ^{n+1}. (Note then that τ^{n+1} cannot be the face of any simplex in K.) A triangulated manifold has a free face if and only if it has a boundary. Because of the existence of a top-dimensional homology class a compact manifold is homeomorphic to a proper subset of itself if and only if it has a boundary. A similar argument will not work for arbitrary finite simplicial complexes because there exist contractible complexes with no free faces (e.g. the Dunce's Hat [Z]). The statement in the title of this paper answers a question first raised by Paul Dierker in his unpublished 1967 Ph.D. thesis while investigating collapsibility of contractible complexes. The question was communicated to the author by David Bellamy. The author thanks Marshall Cohen for pleasant conversations that resulted in a clearer exposition of the proof.

THEOREM: *Suppose* K *is a finite simplicial complex. Then* K *has a free face if and only if* K *is homeomorphic to a proper subset of itself.*

The "only if" part is clear. Before proving the "if" part of the Theorem, we first state some definitions and prove a Lemma. Let n = dimension (K) and define M to be the set of all x in K which have a neighborhood whose closure, $N(x)$, is homeomorphic to an n-ball. Define BdM to be the set of all $x \in M$ such that x is on the boundary S^{n-1} of

the n-ball $N(x)$. Now suppose that $h : K \to K$ is a homeomorphism onto a proper subset $h(K) \subset K$. Let $\overline{M}$ denote the
point set closure of M in K. Then $h(\overline{M}) \subset \overline{M}$, because
$K - M$ is at most $(n-1)$-dimensional and thus $h(N(x))$ cannot
be contained in $K - M$.

LEMMA: *If* BdM *is empty, then* $h(\overline{M} - M) \cap M$ *is empty.*

Proof of Lemma. Let C be a component of M and sup-
pose on the contrary that, for some $y \in (\overline{M} - M)$, $x = h(y) \in C$.
Then $C - h(\overline{M})$ is non-empty, because if empty then
$h^{-1}(N(x))$ is a closed neighborhood of y which is impossible
since $y \notin M$. By Invariance of Domain ([H-W], p. 96)
$h(M) \cap C$ is open, since we suppose BdM to be empty. Therefore $h(\overline{M} - M) \cap C$ separates C (into $C - h(\overline{M})$ and
$C \cap h(M)$) and thus is $(n-1)$-dimensional ([H-W], VI. 11).
Let σ be an $(n-1)$-simplex whose simplicial interior $\overset{\circ}{\sigma}$
intersects $h^{-1}(h(\overline{M} - M) \cap C) = (\overline{M} - M) \cap h^{-1}(C)$. Suppose
$y \in \overset{\circ}{\sigma}$. If σ were the face of more than two n-simplexes in
K, then h could not map a neighborhood of y into C (by
Invariance of Domain). If σ were the face of exactly two
n-simplexes of K, then $\overset{\circ}{\sigma} \subset M$. If σ were the face of no
n-simplex of K, then $\overset{\circ}{\sigma} \not\subset \overline{M} - M$. Therefore σ is the face
of precisely one n-simplex of K and thus $\sigma \subset$ BdM which
contradicts our assumption.

Proof of Theorem. If BdM is non-empty, then BdM contains a free face of K. If BdM is empty, then by the Lemma
we have that $h(\overline{M} - M) \cap M$ is empty. Thus, if C_1 and C_2
are any two components of M, then $h(C_1) \cap C_2 = h(\overline{C_1}) \cap C_2$
being both open and closed is either empty or all of C_2.
Since there are only finitely many components, $h(M) \supset M$ and

242   *David W. Henderson*

thus h(K -M)   is a proper subset of K -M, which is a
finite simplicial complex of dimension less than n.

We now finish the proof by proceeding by induction on the
dimension of K.   To start the induction, note that if n = 1
then K -M is a finite number of vertices (or empty) and
thus has no homeomorphism onto a proper subset of itself.
For   n > 1, we assume that all finite simplicial complexes
of dimension less than n satisfy the Theorem.   Then we conclude from the last paragraph that either BdM   contains a
free face of K or that K -M has a free face σ^k.   Since
the dimension of σ is less than n -1, σ is not the face
of *any* n-simplex.   Therefore, σ is a free face in K.

REFERENCES

[H-W]   Hurewicz and Wallman, <u>Dimension Theory</u>, Princeton
University Press, 1941.

[Z]   Zeeman, E. C., "On the Dunce Hat." <u>Topology</u> 2 (1964),
341-358.

The Alexander Linear Isotopy Theorem in E^3

Michael Starbird

1. <u>Introduction</u>. In [3], Kuiper explores the relationship
between isotopies in the differential category and isotopies
in the PL category. Using that relationship, he states results,
one of which [3], implies the Main Theorem below. The Main
Theorem here is a refinement of the Alexander Isotopy Theorem
for PL homeomorphisms of the 3-cell.

This paper contains a detailed outline for a proof of the
Main Theorem. The proof here contrasts with Kuiper's in the
respect that only PL techniques are employed here.

<u>MAIN THEOREM</u>. (The Alexander Linear Isotopy Theorem in
E^3.) Let B be a PL 3-cell in E^3 and h: B $\rightarrow$ B be a PL
homeomorphism so that h|BdB = id.

Then there is a triangulation T of B and an isotopy
h_t: B $\rightarrow E^3$ (t $\in$ [0,1]) so that h_0 = id, h_1 = h, for each
t, h_t|BdB = id, and for each t, h_t is a PL homeomorphism
which is linear on each simplex of the fixed triangulation T.

Furthermore, the isotopy h_t can be chosen to consist of
a finite number of isotopies performed consecutively, each one
of which consists of moving only one vertex of T and moving
that vertex in a straight line.

The type of isotopy described in the conclusion of the Main
Theorem has a clear physical interpretation. One can think of
the 3-cell B as constructed from a finite number of simple
objects, namely, the simplexes in T. The isotopy h_t, then,
is describing a movement of this 3-cell B so that each of

the tetrahedra which make up B remains a tetrahedron throughout the isotopy.

One should compare the Main Theorem with the standard PL
version of Alexander's Isotopy Theorem. Given the same hypothesis as in the Main Theorem, the standard conclusion is that
there is a PL isotopy H: B × I → B so that $H|B × \{0\}$ = id,
$H|B × \{1\}$ = h, and for each t in [0,1], $H|BdB × \{t\}$ = id.
Implicit in this conclusion is the fact that for each t in
[0,1], there is a triangulation T_t with respect to which
$H|B × \{t\}$ is a linear homeomorphism, i.e. linear on each
simplex of T_t. However, for t ≠ s, the triangulations T_t
and T_s may differ. In fact, in the standard proof of the
standard Alexander Theorem, $T_t ≠ T_s$ for any t, s   except
t = 0, s = 1. In the standard proof, the triangulations T_t
contain simplexes of increasingly smaller diameter as t → 0.
These small simplices in a sense contain all the complexities
of the homeomorphism h and disappear in a limit as t
reaches 0.

The physical interpretation of this process is perhaps not
as appealing as that in the conclusion of the Main Theorem.
However, the proof of the standard Alexander Theorem is short
and applies equally well to all dimensions. The proof outlined
here of the Main Theorem lacks these qualities. In order to
aid the reader in following the proof, §2 contains an outline
of the whole proof with references given to show where in the
remainder of the paper further details are to be found.

DEFINITIONS. Let K be a complex and $h_t: K → E^3$
(t ∈ [0,1]) be a continuous family of embeddings. Then h_t
(t ∈ [0,1]) is a *linear isotopy* if and only if there is a
fixed triangulation T of K so that for every t in

[0,1], h_t is a linear embedding on each simplex in T. If a fixed triangulation T which makes h_t a linear isotopy is to be emphasized, we write h_t: (K,T) → E^3 (t ∈ [0,1]) is a linear isotopy or that h_t is a linear isotopy of (K,T).

A linear isotopy h_t: (K,T) → E^3 (t ∈ [0,1]) is a *simple push* if and only if there is a vertex v in T so that for each vertex w ∈ T, w ≠ v, $h_t(w) = h_0(w)$ (t ∈ [0,1]) and $h_t(v) = (1-t)h_0(v) + th_1(v)$. A *push* is a linear isotopy obtained by performing a finite number of simple pushes sequentially.

The Main Theorem can now be stated more succinctly as follows.

MAIN THEOREM. Let B be a PL 3-cell in E^3 and h: B → B be a PL homeomorphism so that h|Bd B = id.
Then there is a linear isotopy h_t: B → E^3 (t ∈ [0,1]) so that h_0 = id, h_1 = h, and, for each t in [0,1], h_t|Bd B = id.

Furthermore, h_t can be chosen to be a push.

2. Guide to the Proof. Section 3 contains lemmas about compositions and extensions of linear isotopies. These lemmas are general facts about linear isotopies and do not have an independent role in the outline of the proof of the Main Theorem.

In §4 the proof starts in earnest by changing the problem. Here it is proved that the Main Theorem is equivalent to Extension Theorem 4.1 which states: Let σ^3 be a 3-simplex in E^3 and h_t: Bdσ^3 → E^3 (t ∈ [0,1]) be a linear isotopy such that $h_0 = h_1$ = id. Then there is a linear isotopy H_t: σ^3 → E^3 (t ∈ [0,1]) so that for each t in [0,1], H_t|Bd σ^3 = h_t and $H_0 = H_1$ = id.

Rephrased, this Extension Theorem 4.1 states that if the boundary of a 3-simplex in E^3 is moved around for a while and returns to the identity, then the whole 3-simplex can follow along, also returning to the identity.

The principal part of the proof of the Main Theorem is supplying a proof of Extension Theorem 4.1. This is done in §5 through §9. To prove Extension Theorem 4.1, however, the first step is to prove a special case of the Main Theorem. The reader should not consider this special case of the Main Theorem as a first step in the proof of the Main Theorem but, instead, as a first step in the proof of Extension Theorem 4.1.

This special case states that level-preserving (i.e., third coordinate preserving) PL homeomorphisms of a 3-cell can be linearly isotoped to the identity as required. This case is used in §6 to prove the analogous special case of Extension Theorem 4.1, namely, if h_t: Bd $\sigma^3 \to E^3$ $(t \in [0,1])$ is level-preserving, then it can be extended as desired.

In §7, it is shown that Extension Theorem 4.1 is equivalent to another, very similar, Extension Theorem 7.1. This reduction takes care of two problems: one, local complications of h_t: Bd $\sigma^3 \to E^3$ $(t \in [0,1])$ around the vertices and, two, it provides a little leeway in the sense that one need not extend h_t itself, but one can vary h_t a small amount and then extend it. However, as far as the broad outline is concerned, one can still consider Extension Theorem 4.1 as the issue, we can now make a niceness assumption about the h_t of the hypothesis.

In §8, the main strategy of the proof of the Extension Theorem appears. Namely, progressively more complicated h_t's are to be considered. Although the level-preserving case was done earlier, the main measure of complexity has not arrived

until this point. The complexity of h_t is measured much as
the complexity of a PL 2-sphere is measured for purposes of
proving the PL Schoenflies Theorem in E^3. In particular,
§8 contains a proof of the simple closed curve cross section
case. This case is the 0 level of complexity. In §9, more
complicated h_t's are reduced to simple closed curve cross
sectional ones by the insertion of horizontal spanning disks
very much as is done in the proof of the PL Schoenflies
Theorem.

3. <u>Useful Lemmas</u>. This section contains general facts about
linear isotopies which are used throughout the proof. Each
one is easily understood if one thinks about linear isotopies
in a very physical way, namely, taking a finite triangulated
complex in E^3 and moving it around in E^3 so that each
simplex remains a simplex.

<u>USEFUL LEMMA 3.1.</u> Let C be a subcomplex of C^+ and
$h_t: C \to E^n$ ($t \in [0,1]$) be a push or linear isotopy. Let
$H_{(1,t)}: C^+ \to E^n$ ($t \in [0,1]$) be a linear isotopy so that for
each t in $[0,1]$, $H_{(1,t)}|C = h_t$ and let $H_{(2,t)}: C^+ \to E^n$
($t \in [0,1]$) be a linear isotopy so that for each t in
$[0,1]$, $H_{(2,t)}|C = h_1|C$. Then there is a linear isotopy
$H_t: C^+ \to E^n$ ($t \in [0,1]$) so that for each t in $[0,1]$,
$H_t|C = h_t$, $H_0 = H_{(1,0)}$ and $H_1 = H_{(2,1)}$.

<u>USEFUL LEMMA 3.2.</u> Let C be a PL 3-cell in E^3 and f, g
be two PL embeddings of a disk D into C so that $f|Bd\ D$
$= g|Bd\ D$ and $f(D) \cap Bd\ C = f(Bd\ D) = g(D) \cap Bd\ C$. Then there
is a linear isotopy $h_t: D \to E^3$ ($t \in [0,1]$) so that $h_0 = f$,
$h_1 = g$, and for each t in $[0,1]$, $h_t|Bd\ D = f|Bd\ D$ and
$h_t(D) \cap Bd\ C = h_t(Bd\ D)$.

USEFUL LEMMA 3.3.   Let $\{h_{(i,t)}\ (t \in [0,1])\}_{i=1}^{n}$ be a collection of linear isotopies of a complex   C   into E^n so that for each $i = 1,2,\ldots,n-1$, $h_{(i,1)} = h_{(i+1,0)}$.   Then the family of maps obtained by performing first $h_{(1,t)}$ then $h_{(2,t)}$ and so on is also a linear isotopy.

USEFUL LEMMA 3.4.   Let $\{h_{(i,t)}\ (t \in [0,1])\}_{i=1}^{n}$ be a collection of linear isotopies and   C   be a complex in E^n so that $h_{(1,0)} = $ id   and for each   i, $i = 1,2,\ldots,n-1$, $h_{(i+1,t)}$ is a linear isotopy of $h_{(i,1)}(C)$ into E^n so that $h_{(i+1,0)} = $ id.   Then there is a linear isotopy h_t $(t \in [0,1])$ of   C   into E^n so that $h_0 = $ id,   and

$$h_1 = h_{(n,1)} \circ h_{(n-1,1)} \circ \cdots \circ h_{(2,1)} \circ h_{(1,1)}.$$

4.   The Equivalence of the Main Theorem to an Extension Theorem.   In this section we prove that the Alexander Linear Isotopy Theorem in E^3 is equivalent to Extension Theorem 4.1 below.   Theorem 4.1 is then proved in succeeding sections.

EXTENSION THEOREM 4.1.   Let σ^3 be a 3-simplex in E^3 and h_t: Bd $\sigma^3 \to E^3$ $(t \in [0,1])$ be a linear isotopy such that $h_0 = h_1 = $ id.   Then there is a linear isotopy H_t: $\sigma^3 \to E^3$ $(t \in [0,1])$ so that for each   t   in   [0,1], $H_t | $ Bd $\sigma^3 = h_t$ and $H_0 = H_1 = $ id.

Claim.   The Alexander Linear Isotopy Theorem in E^3 is equivalent to Theorem 4.1.

Proof.   We show first that the Main Theorem implies Theorem 4.1.   Let the hypotheses to Theorem 4.1 be given.   Then by [2, Theorem 3.3] there is a push $H_{(1,t)}$: $\sigma^3 \to E^3$ $(t \in [0,1])$ so that $H_{(1,0)} = $ id   for each   t   in   [0,1], $H_{(1,t)} | $ Bd $\sigma^3 = h_t$.

By the Main Theorem there is a push $H_{(2,t)}: \sigma^3 \to E^3$
($t \in [0,1]$) so that $H_{(2,0)} = H_{(1,1)}$, $H_{(2,1)} = $ id, and for
each t in $[0,1]$, $H_{(2,t)} | \text{Bd} \sigma^3 = $ id. By Useful Lemma 3.1,
there is a push H_t of σ^3 which satisfies the conclusion
of Theorem 4.1.

Next we show that Theorem 4.1 implies the Alexander Linear
Isotopy Theorem in E^3. We begin by deriving a corollary of
Theorem 4.1.

COROLLARY 4.2. Let σ^3 be a 3-simplex in E^3, g a linear
homeomorphism of E^3 to itself, and $h_t: \text{Bd}\, \sigma^3 \to E^3$ ($t \in [0,1]$)
a push so that $h_0 = $ id and $h_1 = g | \text{Bd}\, \sigma^3$. Then there is a
push $H_t: \sigma^3 \to E^3$ ($t \in [0,1]$) so that for each t in $[0,1]$,
$H_t | \text{Bd}\, \sigma^3 = h_t$, $H_0 = $ id and $H_1 = g | \sigma^3$.

Proof that Theorem 4.1 implies Corollary 4.2. Let
$g_t: E^3 \to E^3$ ($t \in [0,1]$) be an isotopy of E^3 so that $g_0 = $ id,
$g_1 = g^{-1}$, and for every t in $[0,1]$, g_t is a linear homeomorphism (in the vector space sense) of E^3. Then $g_t h_t$
($t \in [0,1]$) satisfies the hypotheses of Theorem 4.1. Let
H'_t ($t \in [0,1]$) be a linear isotopy guaranteed by Theorem 4.1
which extends $g_t h_t$ ($t \in [0,1]$). Then the linear isotopy
$H_t = g_t^{-1} H'_t$ satisfies the conclusion of Corollary 4.2.

We now proceed to prove that Corollary 4.2 implies the
Alexander Linear Isotopy Theorem in E^3. Let the hypotheses
of the Main Theorem be given. Then there is a triangulation
T_1 of B which is shellable and with respect to which f is
a linear homeomorphism. Let $\sigma_1, \sigma_2, \ldots, \sigma_n$ be a shelling of
T_1. By Useful Lemma 3.2 and [2, Theorem 3.3], there is a push
$h_{(1,t)}: T_1^{(2)} = $ (the 2-skeleton of T_1) $\to E^3$ ($t \in [0,1]$) so

that $h_{(1,0)}$ = id, $h_{(1,1)} | \text{Bd } \sigma_1 = f | \text{Bd } \sigma_1$, and for all t, $h_{(1,t)} | \text{Bd } B$ = id. Again by Useful Lemma 3.2 and [2, Theorem 3.3] there is a push $h_{(2,t)}: T_1^{(2)} \to E^3$ (t $\in$ [0,1]) so that $h_{(2,0)} = h_{(1,1)}$, $h_{(2,1)} | \text{Bd } \sigma^2 = f | \text{Bd } \sigma^2$, and for each t in [0,1], $h_{(2,t)} | \text{Bd } B \cup \text{Bd } \sigma_1 = h_{(2,0)} | \text{Bd } B \cup \text{Bd } \sigma_1$. Continue in this manner to produce pushes $h_{(i,t)}$ (t $\in$ [0,1]) of $T_1^{(2)}$. Performing these pushes sequentially produces a push $h_t: T_1^{(2)} \to E^3$ (t $\in$ [0,1]) such that h_0 = id, $h_1 = f | T_1^{(2)}$ and for each t in [0,1], $h_t | \text{Bd } B$ = id.

For each i (i = 1,2,...,n) $h_t | \text{Bd } \sigma_i$ satisfies the hypotheses of Corollary 2.2. Thus Corollary 4.2 can be used to extend h_t to the interior of each σ_i and thereby prove the Alexander Linear Isotopy Theorem in E^3. The equivalence of the Main Theorem to Extension Theorem 4.1 has thus been proved.

In the succeeding sections, Extension Theorem 4.1 will be proved thereby proving the Main Theorem.

5. **A Special Case of the Main Theorem.** The following Level-Preserving Theorem 5.1 is a special case of the Main Theorem. It will be used later in proving the Extension Theorem 4.1.

LEVEL-PRESERVING THEOREM 5.1. Let B be a convex, polyhedral 3-cell in E^3 and f: B $\to$ B be a PL homeomorphism so that $f | \text{Bd } B$ = id and f is level-preserving, that is, for each point (x,y,z) in B, f(x,y,z) = (x',y',z) and $f | \text{Bd } B$ = id. Then there is a linear isotopy or push $h_t: B \to E^3$ (t $\in$ [0,1]) so that h_0 = id, h_1 = f, and, for each t in [0,1], h_t is level-preserving and $h_t | \text{Bd } B$ = id.

The proof of the Level-Preserving Theorem 5.1 proceeds by first doing the case where the 3-cell B is a prism. The

case where B is a general convex 3-cell follows easily from
this result.

The first lemma below shows how to straighten up a neighborhood of a horizontal level. Later in this section, the levels
are stacked up to prove the prism case.

LEMMA 5.2. Let σ^2 be a 2-simplex in E^2 and $f: \sigma^2 \times I$
$(= [0,1]) \to \sigma^2 \times I$ be a homeomorphism so that if $f|Bd(\sigma^2 \times I)$
$= id$ and so that f is level-preserving, that is, for each
$t \in I$, $f(\sigma^2 \times \{t\}) = \sigma^2 \times \{t\}$. Then for each s in I there
is an $\varepsilon_s > 0$ and a linear isotopy $H_{(s,t)}: \sigma^2 \times I \to E^3$
$(t \in [0,1])$ so that

(1) for each t in $[0,1]$, $H_{(s,t)}|Bd(\sigma^2 \times I) = id$,

(2) $H_{(s,t)}$ is level-preserving for each t,

(3) $H_{(s,t)}|\sigma^2 \times \{s\} = f|\sigma^2 \times \{s\}$,

(4) $H_{(s,0)} = f$, and

(5) $H_{(s,1)}(x,r) = (f(x),r)$ for each $x \in \sigma^2$ and

$r \in (s - \varepsilon_s, s + \varepsilon_s)$.

Proof. (R.H. Bing) Let the hypotheses be given. Let T_s
be a triangulation of $\sigma^2 \times I$ with respect to which f is
linear and containing $\sigma \times \{s\}$ as a subcomplex.

For each 3-simplex τ^3 of T_x which contains a 2-face on
$\sigma \times \{s\}$ pick an interior point (x,r) so that it is possible
to subdivide τ^3 by coning from (x,r) to $Bd \tau^3$ and then
move $f(x,r)$ to the point whose first coordinate equals the
first coordinate of $f(x,s)$ and second coordinate is r by
moving $f(x,r)$ in a straight line and coning out to $Bd(f(\tau^3))$.
Performing such a move on each such 3-simplex τ results a new
triangulation T_s^1 which is a subdivision of T_s and a new

linear homeomorphism of $\sigma^2 \times I$ with respect to T_s^1 which agrees with f on $\sigma^2 \times \{s\}$ and is obtained from f by a linear isotopy satisfying properties (1) – (4) of the conclusion.

For each 1-simplex τ in T_s which lies in $\sigma^2 \times \{s\}$ consider the set of all 3-simplexes in T_s^1 whose intersection with $\sigma^2 \times \{s\}$ is τ. These 3-simplexes form a fan above and below τ. By subdividing them close to the $\sigma^2 \times \{s\}$ level, they can be moved from the position they now lie in to a position which satisfies conclusion (5) on them by a linear isotopy which satisfies conclusions (1) – (4). What remain in the wrong position are 3-simplexes which meet $\sigma^2 \times \{s\}$ at a single vertex. For each vertex v in $\sigma^2 \times \{s\}$. Consider the set of 3-simplexes which intersect $\sigma^2 \times \{s\}$ only at v by choosing a level close to the s level, these can be seen to contain a cone from v over a disk at a higher level. Use [2, Theorem 5.2] to push this disk to where it should be to satisfy conclusion (5) coning from v and extending to a linear isotopy satisfying conditions (1) – (4) and not undoing any of the good previously done. Performing such linear isotopies for each vertex v in $\sigma^2 \times \{s\}$ causes an interval of size $\varepsilon_s > 0$ to be moved to a position required by the conclusion.

The following lemma is the special case of the Level-Preserving Theorem in which the 3-cell considered is a prism, i.e., $\sigma^2 \times I$. For convenience, the linear isotopy in the conclusion goes from f to the identity rather than the reverse.

PRISM CASE LEMMA 3.3. Let σ^2 be a 2-simplex in E^2 and $f: \sigma^2 \times I \to \sigma^2 \times I$ be a PL level-preserving homeomorphism fixed on $Bd(\sigma^2 \times I)$. Then there is a level-preserving linear isotopy $h_t \colon \sigma^2 \times I \to E^3$ $(t \in [0,1])$ so that $h_0 = f$, $h_1 = $ id, and for each t in $[0,1]$, $h_t | \mathrm{Bd}(\sigma^2 \times I) = $ id.

Proof. For each s in $[0,1]$ obtain an $\varepsilon_s > 0$ and linear isotopy $H_{(s,t)}$ of $\sigma^2 \times I$ satisfying the conclusions of Lemma 3.1. Using the compactness of $[0,1]$ find points $\{s_i\}_{i=0}^n$ in $[0,1]$ so that $s_0 = 0 < s_1 < s_2 < \cdots < s_n = 1$ and so that $[0,1] \subset \bigcup_{i=0}^n (s_i - \varepsilon_{s_i}, s_i + \varepsilon_{s_i})$. Next find points $\{t_i\}_{i=0}^n$ in $[0,1]$ so that for each i $(i = 0,1,\ldots,n-1)$, $t_i \in (s_i - \varepsilon_{s_i}, s_i + \varepsilon_{s_i}) \cap (s_{i+1} - \varepsilon_{s_{i+1}}, s_i + \varepsilon_{s_{i+1}})$ and $t_n = s_n = 1$.

The linear isotopy h_t we seek will be obtained by performing several linear isotopies sequentially.

The first. First perform $H_{(s_0=0,t)}$ $(t \in [0,1])$. This linear isotopy pushes $\sigma^2 \times [0,t_0]$ back to the identity.

The second. Next perform the linear isotopy $g_{(2,t)}$ $(t \in [0,1])$ obtained by performing $H_{(0,(1-t))}$ $(t \in [0,1])$ on all points in $\sigma^2 \times [t_0,1]$ and extending to $\sigma^2 \times [0,t_0]$ by coning from an interior point v of $\sigma^2 \times [0,s_1]$ to $\mathrm{Bd}(\sigma^2 \times [0,t_0])$. Note that this linear isotopy $g_{(2,t)}$ is level-preserving, leaves $\mathrm{Bd}(\sigma^2 \times I)$ fixed, and that $g_{(2,1)} | \sigma^2 \times [t_0,1] = f | \sigma^2 \times [t_0,1]$.

The third. Next perform the linear isotopy $g_{(3,t)}$ $(t \in [0,1])$ which is defined by $g_{(3,t)} | \sigma^2 \times [t_0,1] = H_{(s_1,t)} | \sigma^2 \times [t_0,1]$ and $g_{(3,t)} | \sigma^2 \times [0,t_0]$ is obtained by coning from v to $\mathrm{Bd}(\sigma^2 \times [0,t_0])$.

The fourth. Next perform the level-preserving linear isotopy $g_{(4,t)}$ $(t \in [0,1])$ which is defined by

$g_{(4,t)}|\sigma^2 \times [t_1,1] = H_{(s_1,1-t)}|\sigma^2 \times [t_1,1]$. For each $x \in \sigma^2$ and $r \in [t_0,t_1]$, the first coordinate of $g_{(4,t)}(x,r)$ equals the first coordinate of $H_{(s_1,1-t)}(x,t_1)$ and the second remains r. Finally, $g_{(4,t)}|\sigma^2 \times [0,t_0]$ is again obtained by coning from v to $Bd(\sigma^2 \times [0,t_0])$.

The homeomorphism $g_{(4,1)}$, agrees with f on $\sigma^2 \times [t_1,1]$, has lined the fibers of $\sigma^2 \times [t_0,t_1]$ straight under $g_{(4,1)}(\sigma^2 \times \{t_1\})$, and is determined on $\sigma^2 \times [0,t_0]$ by coning from v to $Bd(\sigma^2 \times [0,t_0])$.

The remaining ones. We proceed up through the t_i's straightening out more and more as we go. For each i there is a two-step process. Suppose we have performed linear isotopies so that at the conclusion, $\sigma^2 \times [t_{i-1},1]$ is mapped as f maps it, $\sigma^2 \times [t_0,t_{i-1}]$ is mapped as a product under $\sigma^2 \times \{t_{i-1}\}$ and the homeomorphism is completed to $\sigma^2 \times [0,t_0]$ by coning from v . to $Bd(\sigma^2 \times [0,t_0])$. We proceed to show how to straighten up the map up to t_i.

First perform the linear isotopy g_t' ($t \in [0,1]$) defined by $g_t'|\sigma^2 \times [t_{i-1},1] = H_{(s_i,t)}|\sigma^2 \times [t_{i-1},1]$. For each $x \in \sigma^2$ and $r \in [t_0,t_{i-1}]$, the first coordinate of $g_t'(x,r)$ equals the first coordinate of $H_{(s_i,t)}(x,t_{i-1})$ and the second coordinate of $g_t'(x,r)$ equals r. Finally, $g_t'|\sigma^2 \times [0,t_0]$ is obtained by coning from v out to $Bd(\sigma^2 \times [0,t_0])$ as usual.

Next we define a linear isotopy g_t''. Let $g_t''|\sigma^2 \times [t_i,1] = H_{(s_i,1-t)}|\sigma^2 \times [t_i,1]$; for each $x \in \sigma^2$ and $r \in [t_0,t_i]$ let the first coordinate of $g_t''(x,r)$ equal the first coordinate of $H_{(s_i,1-t)}(x,t_i)$ and the second coordinate equal r, finally, as always, let $g_t''|\sigma^2 \times [0,t_0]$ be obtained by coning from v to $Bd(\sigma^2 \times [0,t_0])$.

By this method we work up through the t_i's. Note that when i = n, $g_1'' = $ id (actually $g_1' = $ id) and the lemma is proved.

Finally, in this section we conclude that the convex case follows from the prism case above.

Proof of Theorem 5.1. Consider the highest and lowest horizontal levels which contain points of B. Then the techniques of Lemma 3.2 can be used to obtain a level-preserving linear isotopy fixed on Bd B so that a neighborhood of both such extreme levels is moved to the desired position. The remainder of the 3-cell is homeomorphic to a prism via a level-preserving PL homeomorphism g. Lemmas 3.2 and 3.3 should then be modified so that the vertical lines of those lemmas are replaced by the lines given using the PL homeomorphism g. The proofs of these modified lemmas are the same as the proofs of Lemmas 3.2 and 3.3, so the convex case is proved.

6. The Level-Preserving Case of Extension Theorem 4.1. Level-Preserving Theorem 5.1 can be used to prove a level-preserving version of Extension Theorem 4.1.

THEOREM 6.1. Let σ^3 be a 3-simplex in E^3 and h_t: Bd $\sigma^3 \to E^3 = E^2 \times E^1$ ($t \in [0,1]$) be a linear isotopy such that $h_0 = h_1 = $ id and h_t is level-preserving in the sense that for each point (x,r) in Bd σ^3 and t in $[0,1]$, $h_t(x,r) = (y,r)$. Then there is a linear isotopy $H_t : \sigma^3 \to E^3$ ($t \in [0,1]$) so that for each t in $[0,1]$, $H_t |$ Bd $\sigma^3 = h_t$ and $H_0 = H_1 = $ id. Furthermore, H_t can be chosen to be level-preserving.

Proof. The first step is to find a linear isotopy $H_{(1,t)} : \sigma^3 \to E^3$ ($t \in [0,1]$) so that for each t in $[0,1]$, $H_{(1,t)} |$ Bd $\sigma^3 = h_t$, $H_{(1,t)}$ is level-preserving, and $H_{(1,0)} = $ id. One can obtain the linear isotopy $H_{(1,t)}$ by noting

that the extension produced in the proof of [2, Theorem 3.3]
is in fact level-preserving. The linear isotopy $H_{(1,t)}$
$(t \in [0,1])$ fails to satisfy the conclusion of Theorem 6.1
only in that $H_{(1,1)}$ may not equal $H_{(1,0)}$. However, since
$H_{(1,1)}$ is a level-preserving PL homeomorphism of σ^3 fixed
on Bd σ^3, Level-Preserving Theorem 5.1 can be used to produce another linear isotopy $H_{(2,t)}$ $(t \in [0,1])$ which begins
at $H_{(1,1)}$, ends at the identity and leaves Bd σ^3 fixed
throughout. The existence of $H_{(1,t)}$ and $H_{(2,t)}$ allows us
to conclude, using Useful Lemma 3.1, that Theorem 6.1 is true.

7. **An Easier, But Equivalent, Extension Theorem.** In dealing
with cases of Extension Theorem 4.1 which are more general than
the level-preserving cases considered already, some problems
arise as a result of complications around vertices. In this
section we show how to avoid many of these problems by adding
a nice collar to Bd σ^3. In particular, we show that Extension
Theorem 7.1 below is equivalent to Extension Theorem 4.1.

   Definitions: 1. A PL 2-sphere Σ in E^3 is *nicely
collarable* if and only if there is a PL 2-sphere Σ' in
Int Σ which is "parallel" to Σ. Here "parallel" means that
there is an $\varepsilon > 0$ and a linear homeomorphism h: $\Sigma \rightarrow \Sigma'$ so
that for each 2-simplex σ of Σ, $h(\sigma)$ lies in a plane parallel to the plane spanned by σ and at distance ε from it.

   2. A linear isotopy h_t: Bd $\sigma^3 \rightarrow E^3$ $(t \in [0,1])$ is a *nice
linear isotopy* if and only if for each t, $h_t(\text{Bd } \sigma^3)$ is
nicely collarable.

   EXTENSION THEOREM 7.1. Let σ^3 be a 3-simplex in E^3,
h_t: (Bd σ^3,T) $\rightarrow E^3$ $(t \in [0,1])$ be a nice linear isotopy such

that $h_0 = h_1 = $ id, and $\mu > 0$. Then there is a linear isotopy $H_t : \sigma^3 \to E^3$ ($t \in [0,1]$) such that $H_t | $Bd σ^3 is linear with respect to T, for each t in $[0,1]$, $d(h_t, H_t | $Bd $\sigma^3) < \mu$, and $H_0 = H_1 = $ id.

We do not prove Extension Theorem 7.1 here. Instead, we prove its equivalence to Extension Theorem 4.1 and hence its equivalence to the Main Theorem.

Claim. Extension Theorem 7.1 is equivalent to Extension Theorem 4.1.

Proof of Claim. We only need to show that the seemingly weaker Extension Theorem 7.1 implies Extension Theorem 4.1. Let the hypothesis of Extension Theorem 4.1 be given. Then h_t: Bd $\sigma^3 \to E^3$ ($t \in [0,1]$) is a linear isotopy with $h_0 = h_1 = $ id.

We use h_t: Bd $\sigma^3 \to E^3$ ($t \in [0,1]$) to produce a linear isotopy h_t': Bd $\sigma^3 \times [0,1] \to E^3$ ($t \in [-1,2]$) such that:

(1) $h_{-1}' = h_2'$ is the neatest possible embedding of an interior collar on Bd σ^3,

(2) $h_t' | $Bd $\sigma^3 \times \{0\} = $ id for $t \in [-1,0] \cup [1,2]$,

(3) $h_t' | $Bd $\sigma^3 \times \{0\} = h_t$ for $t \in [0,1]$ and

(4) $h_t' | $Bd $\sigma^3 \times \{1\}$ ($t \in [-1,2]$) is a nice linear isotopy.

In other words h_t' is a linear isotopy of a collar whose interior boundary is nicely collarable at all times. Notice that after h_t' is produced, it proves that Extension Theorem 7.1 implies Extension Theorem 4.1. This is true because the extension of h_t of Extension Theorem 4.1 can be accomplished in two pieces: first, use h'_t to extend h_t to a collar; and, second, use the conclusion of Extension Theorem 7.1 to complete the extension of h_t to all of σ^3.

We proceed to describe h'_t: Bd $\sigma^3 \times [0,1] \to E^3$ ($t \in [-1,2]$). To describe h'_t we will first describe a triangulation S of Bd $\sigma^3 \times [0,1]$ with respect to which h'_t is linear and then explain how the vertices move for $t \in [-1,2]$.

All vertices of the triangulation S of Bd $\sigma^3 \times [0,1]$ lie on the two boundary components. Let T be a triangulation of Bd σ^3 with respect to which h_t ($t \in [0,1]$) is linear. The vertices of S on Bd $\sigma^3 \times \{0\}$ are simply the vertices of T. It remains to describe the vertices of S which lie on Bd $\sigma^3 \times \{1\}$. (See Figure 1.)

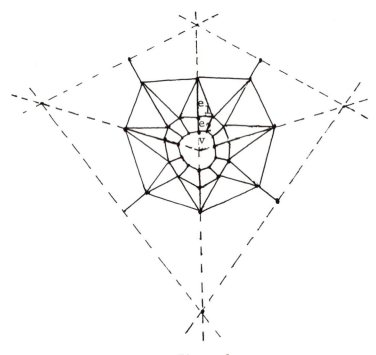

Figure 1

The triangulation of $S \mid Bd \ \sigma^3 \times \{1\}$ is obtained by considering the triangulation T. The dotted lines represent the triangulation T. Each vertex v is surrounded by three concentric simple closed curves in the 1-skeleton of S. The outer curve has as its vertices the barycenters of the 2-simplices and 1-simplices in $St(v,T)$. There are twice as many simplices in that simple closed curve as there are 2-simplexes in $St(v,T)$. The next simple closed curve inward has three halves as many vertices as the above curve and is pictured. Let D_v be the disk bounded by this curve. The inner ring is parallel to the second. The trapezoids pictured can be divided in half to obtain a triangulation. Finally the remaining part of the disk D_v around v, will be triangulated as required later.

The linear isotopy h'_t ($t \in [-1,2]$) will be described in three pieces, namely, $t \in [-1,0]$, $t \in [0,1]$, and $t \in [1,2]$. The parts $t \in [-1,0]$ and $t \in [1,2]$ are simply transitional parts taking us from the identity embedding of $Bd \ \sigma^3 \times [0,1]$ to h'_0 and from h'_1 back to the identity. The interesting piece is $t \in [0,1]$.

For each vertex u of S on $Bd \ \sigma^3 \times \{1\}$, there is one (or more) 2-simplex $\sigma(u)$ in T to which u corresponds in the sense that if $Bd \ \sigma^3 \times \{1\}$ were projected to $Bd \ \sigma^3$, u would lie in $\sigma(u)$. Let u be a vertex of S on $Bd(\sigma^3 \times \{1\})$ which is not interior to any D_v. We define h'_t ($t \in [0,1]$) in such a way that there is a fixed $\delta > 0$ with the following property. For each $t \in [0,1]$, u, a vertex of S as above, and any $\sigma(u)$, $h'_t(u)$ lies in the plane which is parallel to the plane determined by $\sigma(u)$ and at distance δ from it. This implies that for each time $t \in [0,1]$, the vertices as above which lie right over an edge of T are mapped into the bisection of the dihedral angle formed by $h_t(\sigma_1)$ and $h_t(\sigma_2)$

where σ_1 and σ_2 are the two 2-simplexes which contain the edge.

Let $\varepsilon > 0$ be a number so small that if v, w are vertices of T, $d(h_t(v), h_t(w)) > 3\varepsilon$ for each $t \in [0,1]$. For each vertex v of T and t in $[0,1]$, let $B(v,t)$ be the round ball of radius ε centered at $h_t(v)$.

For each $t \in [0,1]$, v, a vertex in T, and vertex u of D_v, $h_t'(u)$ lies on $Bd\ B(v,t)$. The vertices of S which lie on the edges of T will be mapped into the bisector of the appropriate dihedral angle determined by images of two 2-simplices of T. The vertices of $Bd\ D_v$ and its collar are mapped by h_t' carefully so that $h_t'(Bd\ \sigma^3 \times \{1\})$ will be nicely collarable. The vertices in $h_t'(Int\ D_v)$ will be nicely collarable since they lie on a round 2-sphere $Bd(B(v,t))$. The danger to nice collarability occurs at the image of $Bd\ D_v$. The triangulation of $Bd\ D_v$ and its collar, however, make it possible to define the map h_t' so it will be nicely collarable. This proves that Extension Theorem 7.1 implies Extension Theorem 4.1.

8. The Simple Closed Curve Cross Section Case of Extension Theorem 7.1.

Our object now is to take a nice linear isotopy $h_t: Bd\ \sigma^3 \to E^3$ ($t \in [0,1]$) such that $h_0 = h_1 = id$ and extend an approximation of h_t to $H_t: \sigma^3 \to E^3$ ($t \in [0,1]$) such that $H_0 = H_1 = id$. The plan is to start with simple isotopies h_t and work up to more complicated ones. The measure of simplicity of h_t is similar to that used in the standard proof of the PL Schoenflies Theorem in E^3. In this section we consider the simplest case

THEOREM 8.1. Let $h_t: Bd\ \sigma^3 \to E^3 = E^2 \times E^1$ ($t \in [0,1]$) be a nice linear isotopy such that

(1) $h_0 = h_1 = $ id   and

(2)   for each $t \in [0,1]$ and $s \in E^1$,
      $(E^2 \times \{s\}) \cap h_t(Bd \ \sigma^3)$ is empty, a simple closed
      curve, or s is the highest or lowest level that
      $h_t(Bd \ \sigma^3)$ intersects.

Then there is a linear isotopy $H_t : \sigma^3 \to E^3$ $(t \in [0,1])$
such that $H_t | Bd \ \sigma^3 = h_t$ $(t \in [0,1])$ and $H_0 = H_1 = $ id.

   Proof.   The fact that h_t is a nice linear isotopy implies
that there are an $\varepsilon > 0$ and a linear isotopy
g_t: $Bd \ \sigma^3 \times [0,\varepsilon] \to E^3$ $(t \in [0,1])$ such that for each
$t \in [0,1]$, $g_t(Bd \ \sigma^3 \times \{\delta\})$ is parallel to $h_t(Bd \ \sigma^3)$ at
distance δ from it.   In other words, the collar on $h_t(Bd \ \sigma^3)$
is a very neat one which has natural levels to it.
   The proof of Theorem 8.1 has two steps.
   Step 1.   There is a linear isotopy g_t': $Bd \ \sigma^3 \times [0,\varepsilon] \to E^3$
$(t \in [0,1])$ such that

   (1)   for each $t \in [0,1]$ $g_t' | Bd \ \sigma^3 \times \{0\} = h_t$,

   (2) $g_0' | Bd \ \sigma^3 \times [0,\varepsilon] = $ identity,

   (3) $g_1' | Bd \ \sigma^3 \times \{\varepsilon\} = $ identity, and

   (4)   for each $\delta \in [0,\varepsilon]$, $g_1'(Bd \ \sigma^3 \times \{\delta\}) = $
         $g_1(Bd \ \sigma^3 \times \{\delta\})$.

   The last condition says that g_1' brings the levels of the
collar set-wise where they should go, but not necessarily
pointwise.
   The second step completes the process of bringing the interior of σ^3 back to the identity.

   Proof of Step 1.   The proof here makes use of g_t to guide
the construction of g_t'.   By stretching and shrinking the E^1

factor of $E^3 = E^2 \times E^1$, we assume that the highest and lowest points of $g_t(\text{Bd } \sigma^3 \times \{1\})$ are always on the same levels. Then one proves the following fact.

For each $t \in [0,1]$, there is a $\delta(t) > 0$ such that if r and $s \in ([0,1] \cap (t - \delta(t), t + \delta(t)))$ (and r, s are not among the finite number of critical times when two vertices are on the same level) then there is a linear isotopy $j_u\colon \text{Bd } \sigma^3 \times [0,\varepsilon] \to E^3$ $(u \in [r,s])$ such that

(1) $j_0 = g_r$,

(2) for each $u \in [r,s]$, $j_u | \text{Bd } \sigma^3 \times \{0\} = h_t$,

(3) for each $u \in [r,s]$, $j_u | \text{Bd}\sigma^3 \times \{1\}$ is level-preserving, and

(4) $j_s(\text{Bd } \sigma^3 \times \{\delta\}) = g_s(\text{Bd } \sigma^3 \times \{\delta\})$.

The last condition again says that the levels of the collar return setwise to their position under g_s.

We omit the proof of the above fact. Suppose that such j_u's exist. Then the intervals $\{(t - \delta(t), t + \delta(t))\}_{t \in [0,1]}$ form an open cover of $[0,1]$. By compactness of $[0,1]$ find a finite subcover in ascending order of t_i's, $\{(t_i - \delta(t_i), t_i + \delta(t_i))\}_{i=1}^{n}$. Find points $0 = r_0 < r_1 < \cdots < r_n = 1$ so that for each $i = 1,\ldots,n-1$ $t_{i+1} - \delta(t_{i+1}) < r_i < t_i + \delta(t_i)$.

Use the j_u's produced above to give linear isotopies which go between consecutive r_i's. Piecing them together gives a linear isotopy of $\text{Bd } \sigma^3 \times [0,1]$ required in Step 1. Note that the linear isotopy so obtained agrees setwise with the guide g_t at the times $\{r_i\}_{i=0}^{n}$. Also note that at all times the image of $\text{Bd } \sigma^3 \times \{\varepsilon\}$ is level-preserving.

After we get to time 1, an additional move must be made to return $\text{Bd } \sigma^3 \times \{\varepsilon\}$ to the identity.

The Level-Preserving Case of Theorem 4.1, namely, Theorem 6.1, is now used to return $\sigma^3 - (\text{Bd } \sigma^3 \times [0,\varepsilon])$ to the identity.

Step 2. Let $f: \sigma^3 \to \sigma^3$ be a PL homeomorphism so that

    (1) $f \mid \text{Bd } \sigma^3 = \text{id}$,

    (2) $f \mid \sigma^3 - (\text{Bd } \sigma^3 \times [0,\varepsilon]) = \text{id}$, and

    (3) for each $\delta \in [0,1]$, $f(\text{Bd } \sigma^3 \times \{\delta\}) = \text{Bd } \sigma^3 \times \{\delta\}$.

Then there is a linear isotopy $F_t: \sigma^3 \to E^3$ ($t \in [0,1]$) so that $F_0 = f$, $F_1 = \text{id}$, and for each $t \in [0,1]$, $F_t \mid \text{Bd } \sigma^3 = \text{id}$.

Proof of Step 2. The proof is to view the collar as a product and use the Level-Preserving Case of the Main Theorem, namely, Theorem 5.1, to straighten it. It is first necessary to prove that there is a linear isotopy of σ^3 which leaves Bd σ^3 fixed and takes f to a homeomorphism f' which has all the virtues that f had and, in addition, f' (the six 1-simplices of σ^3) $\times [0,\varepsilon] = \text{id}$. For each one of the four 2-simplexes of Bd σ^3, say τ, $f' \mid \tau \times [0,\varepsilon]$ is a geometrically level-preserving homeomorphism. Hence Theorem 5.1 proves that each of those four pieces can be returned to the identity.

    This step completes the proof of the level-preserving case of Extension Theorem 7.1.

9. Scheme for the General Case of Extension Theorem 7.1.

Here we indicate how induction on the complexity of h_t can be used to reduce Extension Theorem 7.1 to the simple closed curve cross section case proved in the previous section.

EXTENSION THEOREM 7.1. Let h_t : $(\text{Bd } \sigma^3, T) \to E^3$ $(t \in [0,1])$ be a nice linear isotopy such that $h_0 = h_1 = \text{id}$ and $\mu > 0$. Then there exists a linear isotopy H_t : $\sigma^3 \to E^3$ $(t \in [0,1])$ so that $H_t | \text{Bd } \sigma^3$ is linear with respect to T, for each $t \in [0,1]$, $d(h_t, H_t | \text{Bd } \sigma^3) < \mu$, and $H_0 = H_1 = \text{id}$.

Proof. As in the standard proof of the PL Schoenflies Theorem, we measure the complexity of h_t and reduce it using a θ-curve theorem.

The complexity of a PL 2-sphere Σ is measured similarly to how it is done in the standard proof of the PL Schoenflies Theorem. Here we float a horizontal plane up starting below Σ and watch the intersections with Σ. A critical level contains a vertex and occurs when the intersection of Σ with the floating plane changes topologically. Assume no two vertices occur at the same level. At each critical level count the number of simple closed curves there. Count each loop of bouquets. Take the sum of these numbers over all critical levels.

During the course of a linear isotopy $h_t : \text{Bd } \sigma^3 \to E^3$ $(t \in [0,1])$, there are critical times at which two or more vertices lie at the same level. We have enough leeway in our choice of h_t to assume that there are only a finite number of critical times and then exactly two vertices are at the same level. Suppose $0 = t_0 < t_1 < t_2 < \cdots < t_n = 1$ are the critical times. Then for each $s, t \in (t_i, t_{i+1})$, $0 \leq i < n$, $h_s(\text{Bd } \sigma^3)$ and $h_t(\text{Bd } \sigma^3)$ have the same complexity. Also, there are some critical times which are not important because the complexity does not change as a result. Ignore those critical times.

The complexity of $h_t : \text{Bd } \sigma^3 \to E^3$ $(t \in [0,1])$ is measured as a pair (k,m) where k is the maximal complexity of

$h_t(Bd\ \sigma^3)$ and   m   is the number of intervals between important critical times during which the maximum is reached.

Our induction is on the pair   (k,m)   lexicographically
ordered.   In §8, we did the   k = 0   case.

Since h_t is a nice linear isotopy, it can be extended to
a linear isotopy of a nice collar say h'_t: Bd σ^3 × [0,1] → E^3
(t ∈ [0,1]).   The map $h'_t|$Bd σ^3 ×{1}   has the same complexity
as h_t.   We will use the collar to accommodate the same type
of slipping as occurred in Step 1 of Theorem 8.1.

Suppose the maximal complexity of $h'_t|$Bd σ^3 ×{1}   occurs
between important critical times   r   and   s.   Then there is a
PL isotopy j_t: S^1 → E^3 (t ∈ (r,s))   so that

(1) $j_t(S^1) \subset h'_t(Bd\ \sigma^3 ×\{1\})$,

(2) $j_t(S^1)$ bounds a disk D_t which lies on a horizontal plane and whose interior misses
$h'_t(Bd\ \sigma^3 ×\{1\})$,   and

(3) $j_t(S^1)$ divides $h'_t(Bd\ \sigma^3 ×\{1\})$ into two disks
D_1 and D_2 so that $D_t \cup D_1$ has complexity   0
and $D_2 \cup D_t$ has complexity less than   k.

The inductive proof is completed as follows.   Find a simple
closed curve   J   on   Bd σ^3 × {1},   a modified collar isotopy
h''_t : Bd σ^3 × [0,1] → E^3 (t ∈ [0,1]),   and a disk   D   with
Bd D = J   and   Int D ∩ Bd σ^3 × {1} = φ   so that

(1) $h''_0 = h''_1$ = id,

(2) h''_t (t ∈ [0,1])   from time to time setwise preserves
the levels of the collar h'_t (t ∈ [0,1])   exactly
as was done in Step 1 of Theorem 8.1,

(3)   Let D_1, D_2 be the disks on   Bd σ^3 × {1}   bounded
by   J.   Then there is a linear isotopy

$\tilde{h}_t : (D \cup D_1 \cup D_2) \to E^3$ $(t \in [0,1])$ with $\tilde{h}_0(D \cup D_i)$ $(i = 1,2)$ having complexity 0 and so that for $i = 1,2$, there is a linear isotopy $\tilde{h}_{(i,t)} : (D \cup D_i) \times [0,1] \to E^3$ $(t \in [0,1])$ so that $\tilde{h}_{(i,0)} = \tilde{h}_{(i,1)} = \text{id}$, $\tilde{h}_{(i,t)} | (D \cup D_i) \times \{1\}$ is a nice linear isotopy with complexity less than the complexity of h_t.

If $\text{Int } D \cap h_0''$ ($\text{Bd } \sigma^3 \times [0,1]$) = ϕ, then induction on complexity is easily carried out. Otherwise, the following lemma is used to finish the inductive step.

LEMMA 9.1. Let $B = B_1 \cup B_2$ be a PL 3-cell in E^3 where B_1 (also B_2) is a PL 3-cell in B with $\text{Bd } B_1 \cap \text{Bd } B$ a 2-cell and $B_2 = C\ell(B - B_1)$. Let $h_t : (\text{Bd } B \cup \text{Bd } B_1) \to E^3$ $(t \in [0,1])$, $F_t : B \to E^3$ $(t \in [0,1])$ and $G_t : B_1 \to E^3$ $(t \in [0,1])$ be linear isotopies so that for each $t \in [0,1]$, $F_t | \text{Bd } B = h_t | \text{Bd } B$ and $G_t | \text{Bd } B_1 = h_t | \text{Bd } B_1$, and $F_0 = F_1 = \text{id}$ and $G_0 = G_1 = \text{id}$.

Then there exists a linear isotopy $H_t : B_2 \to E^3$ $(t \in [0,1])$ so that $H_t | \text{Bd } B_2 = h_t | \text{Bd } B_2$ and $H_0 = H_1 = \text{id}$.

Proof. Let $D = B_1 \cap B_2$. Let C be a neat PL collar on D in B_2 tapered to $\text{Bd } D$ and let $G_t^+ : B_1 \cup C \to E^3$ $(t \in [0,1])$ be a linear isotopy extending G_t so that $G_0^+ = G_1^+ = \text{id}$.

Let $f : B_1 \cup C \to C$ be a PL homeomorphism so that $f | C\ell(\text{Bd } C - D) = \text{id}$. Consider the isotopy $\tilde{H}_t$ $(t \in [0,1])$ of B_2 defined as follows for each $x \in B_2$:

$$\tilde{H}_t(x) = \begin{cases} G_t^+ \circ f \circ G_t^{+-1} \circ F_t \circ f^{-1} & \text{if } x \in C \text{ and} \\ \qquad\qquad F_t \circ f^{-1}(x) \in G_t^+(B_1) \\[2ex] G_t^+ \circ f \circ G_t^{+-1} \circ F_t & \text{if } x \notin C \text{ and} \\ \qquad\qquad F_t \circ f^{-1}(x) \in G_t^+(B_1) \\[2ex] F_t \circ f^{-1} & \text{if } x \in C \text{ and} \\ \qquad\qquad F_t \circ f^{-1}(x) \notin G_t^+(B_1) \\[2ex] F_t & \text{if } x \notin C \text{ and} \\ \qquad\qquad F_t \circ f^{-1}(x) \notin G_t^+(B_1) \end{cases}$$

The isotopy $\tilde{H}_t$ is not a linear isotopy; however it does have the properties that $\tilde{H}_0 = \tilde{H}_1 = \text{id}$ and $\tilde{H}_t | \text{Bd } B_2 = h_t | \text{Bd } B_2$. The desired linear isotopy H_t is obtained by proving that since $\tilde{H}_t$ is basically a composition of linear isotopies, $\tilde{H}_t$ can be approximated by a linear isotopy which retains the desired properties.

REFERENCES

1. J. W. Alexander, "On the deformation of an n-cell," Proc. Nat. Acad. of Sci. 9 (1923), 406-407.

2. R. H. Bing and M. Starbird, "Linear isotopies in E^2," Trans. Amer. Math. Soc. 237 (1978), 205-222.

3. N. H. Kuiper, "On the smoothings of triangulated and combinatorial manifolds," Diff. and Comb. Topology, Princeton University Press, Princeton, N.J., 1965.